SOCIAL ORIGINS OF Mental Ability

his book fills an urgent, long-felt need.
principle, everyone agrees that intelli-
nt behavior depends heavily on motiva-
n, cultural experience, and acquired
gnitive skill. But in practice no one has
ken the crucial next step, which is to
rvey the research literature in those
mains and see how it adds up....While
e range of topics Collier addresses here
admirable in its own right, his success
bringing them together is even more
....When we finally reach the last chap-
:...we are conscious of having traversed a
ry large and significant intellectual ter-
in."

–from the Foreword by Ulric Neisser

ver the past century many influential
oks and articles have appeared in which
thors have offered "irrefutable" empiri-
l evidence for the genetic origins of
man intelligence. At the same time,
fortunately, nearly all that has been
ritten in defense of the nurture side of
e "nature vs. nurture" debate has been
lemical in nature, concentrating mainly
shooting holes in the opposition's argu-
ents. Perhaps, then, Gary Collier's most
tstanding achievement in authoring this
fense of the social origins of mental abil-
y is in offering a viable synthesis of sup-
orting facts and ideas from the worlds of
cial psychology, the psychology of per-
nality, and cognitive psychology. In so
oing, he has done much to advance the
urture side of the debate.

ocial Origins of Mental Ability is divided
to four interrelated parts. Part One pro-
des a general overview within which the
thor examines some of the major contro-
ersies informing contemporary intelli-
ence research. Some of these include
ebates about the nature and measurement
f intelligence and the relative contribu-
on of genetics and the environment.
ecent research into artificial intelligence
also discussed with particular attention
eing paid to the limitations of the use of
omputer models in the investigation of
uman intelligence.

Part Tw… issues of the psychology of personality, including achievement motivation, locus of control, level of aspiration, and intrinsic motivation. Related areas, such as the fear of success, learned helplessness, resilient children, and self-handicapping strategies are also discussed. Gender differences are stressed in the chapter devoted to achievement motivation, while the differences between blacks and whites are explored in discussions of locus of control, level of aspiration, and self-esteem.

Part Three reviews the research on the development of cognitive skills, beginning with the relationship between language and thought, and covering perception, memory, creative problem solving, and formal thought. It is argued that social conditions and previous experience shape every aspect of mental development, including the speed, breadth, and depth of human information processing.

In the fourth and final part of *Social Origins of Mental Ability,* the author examines the status of blacks in America and reviews the research on early childhood intervention and education reform. The book concludes with a general discussion of the social cost of nonintervention and what may happen if politicians and educators choose to do nothing.

The first comprehensive, systematic survey of research into the nonhereditary influences on intelligence, this book's impact will be felt beyond academe and the psychological community and is certain to have a profound influence upon the thinking of educators and policymakers in the years ahead.

About the author

GARY COLLIER, PhD, is a professor in the Department of Psychology, University College of Cape Breton, Nova Scotia. He has been a visiting scholar at the Emory Cognition Project conducted at the Department of Psychology of Emory University, as well as an invited researcher at the Laboratoire de Psychologie Sociale at the Université de Paris. His other books include *Currents of Thought in American Social Psychology* (coauthored with H. Minton and G. Reynolds) and *Emotional Expression.* Dr. Collier received his doctorate in psychology from the University of Alberta.

SOCIAL ORIGINS OF MENTAL ABILITY

Recent titles in the

Wiley Series on Personality Processes

Irving B. Weiner, *Editor*
University of South Florida

Social Origins of Mental Ability *by Gary Collier*
Symptoms of Schizophrenia *edited by Charles G. Costello*
The Rorschach: A Comprehensive System. Volume I: Basic Foundations (Third Edition) *by John E. Exner, Jr.*
Symptoms of Depression *edited by Charles G. Costello*
Handbook of Clinical Research and Practice with Adolescents *edited by Patrick H. Tolan and Bertram J. Cohler*
Internalizing Disorders in Children and Adolescents *edited by William M. Reynolds*
Assessment of Family Violence: A Clinical and Legal Sourcebook *edited by Robert T. Ammerman and Michel Hersen*
Handbook of Clinical Child Psychology (Second Edition) *edited by C. Eugene Walker and Michael C. Roberts*
Handbook of Clinical Behavior Therapy (Second Edition) *edited by Samuel M. Turner, Karen S. Calhoun, and Henry E. Adams*
Psychological Disturbance in Adolescence (Second Edition) *by Irving B. Weiner*
Prevention of Child Maltreatment: Development and Ecological Perspectives *edited by Diane J. Willis, E. Wayne Holden, and Mindy Rosenberg*
Interventions for Children of Divorce: Custody, Access, and Psychotherapy *by William F. Hodges*
The Play Therapy Primer: An Integration of Theories and Techniques *by Kevin John O'Connor*
Adult Psychopatholody and Diagnosis (Second Edition) *edited by Michel Hersen and Samuel L. Turner*
The Rorschach: A Comprehensive System. Volume II: Interpretation (Second Edition) *by John E. Exner, Jr.*
Play Diagnosis and Assessment *edited by Charles E. Schaefer, Karen Gitlin, and Alice Sandgrund*
Acquaintance Rape: The Hidden Crime *edited by Andrea Parrot and Laurie Bechhofer*
The Psychological Examination of the Child *by Theodore H. Blau*
Depressive Disorders: Facts, Theories, and Treatment Methods *by Benjamin B. Wolman, Editor, and George Stricker, Co-Editor*
Social Support: An Interactional View *edited by Barbara R. Sarason, Irwin G. Sarason, and Gregory R. Pierce*
Toward a New Personology: An Evolutionary Model *by Theodore Millon*
Treatment of Family Violence: A Sourcebook *edited by Robert T. Ammerman and Michel Hersen*
Handbook of Comparative Treatments for Adult Disorders *edited by Alan S. Bellack and Michel Hersen*
Managing Attention Disorders in Children: A Guide for Practitioners *by Sam Goldstein and Michael Goldstein*
Understanding and Treating Depressed Adolescents and Their Families *by Gerald D. Oster and Janice E. Caro*
The Psychosocial Worlds of the Adolescent: Public and Private *by Vivian Center Seltzer*
Handbook of Parent Training: Parents as Co-Therapists for Children's Behavior Problems *edited by Charles E. Schaefer and James M. Briesmeister*
From Ritual to Repertoire: A Cognitive-Developmental Approach with Behavior-Disordered Children *by Arnold Miller and Eileen Eller-Miller*
The Practice of Hypnotism. Volume 2: Applications of Traditional and Semi-Traditional Hypnotism: Non-Traditional Hypnotism *by Andre M. Weitzenhoffer*
The Practice of Hypnotism. Volume 1: Traditional and Semi-Traditional Techniques and Phenomenology *by Andre M. Weitzenhoffer*

Social Origins of Mental Ability

Gary Collier
Professor of Psychology
University College of Cape Breton

A WILEY-INTERSCIENCE PUBLICATION
JOHN WILEY & SONS, INC.
New York • Chichester • Brisbane • Toronto • Singapore

This text is printed on acid-free paper.

Library of Congress Cataloging in Publication Data:

Collier, Gary.
Social origins of mental ability / by Gary Collier.
p. cm. — (Wiley series on personality processes)
Includes bibliographical references and index.
ISBN 0-471-30407-7
1. Nature and nurture. 2. Intellect—Social aspects. I. Title.
II. Series.
BF341.C57 1994
155.9'2—dc20 93-3625

Printed in the United States of America

10 9 8 7 6 5 4 3 2 1

Dedicated to my beloved wife,
Bernadette Gillis Collier

Series Preface

This series of books is addressed to behavioral scientists interested in the nature of human personality. Its scope should prove pertinent to personality theorists and researchers as well as to clinicians concerned with applying an understanding of personality processes to the amelioration of emotional difficulties in living. To this end, the series provides a scholarly integration of theoretical formulations, empirical data, and practical recommendations.

Six major aspects of studying and learning about human personality can be designated: personality theory, personality structure and dynamics, personality development, personality assessment, personality change, and personality adjustment. In exploring these aspects of personality, the books in the series discuss a number of distinct but related subject areas: the nature and implications of various theories of personality; personality characteristics that account for consistencies and variations in human behavior; the emergence of personality processes in children and adolescents; the use of interviewing and testing procedures to evaluate individual differences in personality; efforts to modify personality styles through psychotherapy, counseling, behavior therapy, and other methods of influence; and patterns of abnormal personality functioning that impair individual competence.

IRVING B. WEINER

University of South Florida
Tampa, Florida

Foreword

This book fills an urgent, long-felt need. In principle, everyone agrees that intelligent behavior depends heavily on motivation, cultural experience, and acquired cognitive skill. But in practice, no one has taken the crucial next step, which is to survey the research literature in those domains and see how it adds up. It is easy to see why not: The sheer number of relevant topics is daunting enough. They include not only a host of well-studied individual characteristics—personality traits, cognitive styles, linguistic skills, mnemonic strategies, modes of attention—but also a formidable array of social phenomena: discrimination, segregation, stereotyping, schooling. Who could possibly work through all those different fields carefully enough to assess their implications for intelligence? Gary Collier could, that's who.

Although the range of topics Collier addresses here is admirable in its own right, his success in bringing them together is even more so. For him, the old rubric "personality and social psychology" names one continuous domain, not two independent topics. Whereas many chapters begin with a classical person variable, most of them culminate in a discussion of some related group difference or some policy-related issue (gender and "fear of success," racial segregation, the effects of television on children). The title is meant seriously: This is a book about the *social* origins of mental ability.

Nevertheless, it is also a book about individual differences. Many familiar concepts from the classical psychology of personality—level of aspiration, need for achievement, locus of control, field dependence, the authoritarian personality—are carefully reviewed in these pages, along with others more recently defined, such as fear of success, mastery orientation, learned helplessness. It is a brute fact about American society that such traits are unevenly distributed; most of them show significant correlations with race, sex, and socioeconomic status. If we are ever to understand group differences in apparent mental ability, those correlations cannot be ignored.

Even "personality" and "social psychology" taken together do not exhaust the relevant range of topics here. In Part Three of his book, Collier goes beyond the demands of its title to survey the *cognitive* roots of mental ability; language and its relation to thought; perception, attention, and cognitive style; memory and metamemory; and methods of problem solving and heuristics for creativity. Here, too, the social implications of the research are never out of sight for long. When we finally reach the last chapter, focused on practical issues and unashamedly political in its recommendations, we are conscious of having traversed a very large and significant intellectual terrain.

Years ago, I used to teach a course called "Thought and Intelligence," in which I tried to present the disputed issues of this field to undergraduates as clearly as possible. It seemed to be something they ought to know—indeed, something everyone ought to know. Although the course worked tolerably well, I eventually gave it up. It had come to include too many topics, too incoherently related. Many of those topics lay outside my own expertise, and there was no organized reference source to which I could turn for information. Nor was there any appropriate text for the students, at least none that covered the range of relevant issues in a serious way. Now, at last, there is *Social Origins of Mental Ability.* Courses like mine are possible again, and I will not be the only professor in the United States or Canada who thinks about offering one. As teachers, as researchers, as psychologists, and simply as citizens concerned with social issues, we all owe Gary Collier a vote of thanks for this remarkable book.

ULRIC NEISSER

Emory University
February 1993

Preface

This book attempts to examine the nurture side of the nature–nurture controversy as it pertains to intelligence. Although the importance of the social environment in shaping mental abilities has long been recognized, treatments of its impact have tended to be quite general (for the most part) and this book differs from previous discussions because it attempts to provide a broad overview of the ways that motivation and cultural conditions shape the development and use of cognitive skills. Those who support the environmental position have been far more successful in attacking the genetic position than in building a coherent case of their own.

The book discusses two major areas of concentration. The first focuses on motivation, and the second covers the development of cognitive skills. These two areas are not just different topics, but are different spheres of research, with very little overlap. Although there are many notable exceptions, those who study motivation tend to treat cognition in a very general way and fail to consider the processes and subprocesses responsible for individual differences. Cognitive psychologists, on the other hand, often ignore motivation altogether, apparently assuming that it plays little or no role in information processing or that it can be added to theories of cognition later on. The recent work on helplessness and metacognition may help bridge this gap, bring the two areas together, and serve as a useful model for future research.

Each chapter begins with a brief discussion of the broad-based theoretical position related to the research and includes a general discussion of cultural conditions and cognitive consequences. These intellectual traditions serve as the glue used to hold together the various areas of research within each chapter. The chapter on achievement motivation, for example, begins with a general discussion of Max Weber's concept of the Protestant work ethic. The chapter on intrinsic motivation includes discussions of Gordon Allport's concept of functional autonomy and Abraham Maslow's theory of self-actualization. Several chapters focus on differences between blacks and whites in American society, whereas others stress gender or cross-cultural differences. The overall aim is to bring together a wide range of theory and research drawn from psychology, sociology, anthropology, linguistics, and history. The book is intended primarily for teachers and professionals working in psychology, education, or one of the social sciences; but it could also be used as a supplementary text for graduate or undergraduate courses in cognition, intelligence, or the sociology of knowledge. Because it covers both motivation and cognitive development, it has no predecessors and no direct competition.

The first nine chapters were circulated and discussed in a graduate seminar at Emory University in 1990. I would like to thank the people who participated in this seminar, read parts of the manuscript, and made comments and suggestions. These include Davido Dupree and Professors Jacqueline Irvine, David Jopling, and Ulric Neisser. Parts of the manuscript have also been read by Professors William Clemens, Richard Keshen, and Graham Reynolds at the University College of Cape Breton. Research was funded, in part, by a series of internally administered grants provided by the Social Science and Humanities Research Council of Canada.

GARY COLLIER

Cape Breton
January 1993

Contents

PART ONE

Introduction

CHAPTER 1

The Social Psychology of Intelligence

Traditional research on intelligence can be extended by applying various aspects of social psychology. The social psychological contributions fall into three major areas: (1) Social psychology can help show how specific cultural conditions lead to various forms of adaptive and maladaptive behavior; (2) social psychology has had a long history of research on various aspects of motivation associated with intelligence; (3) social psychology can help uncover some of the biases and distortions occurring during information processing that limit people's ability to solve problems or perceive and recall situations accurately. The chapter ends with a brief discussion of Lev Vygotsky's "zone of proximal development" and its potential role in integrating much of the diverse literature on the social origins of mental ability.

Although research on intelligence[1] has progressed considerably in recent years, there is still a great deal to be done. Part of the problem stems from the fact that the three areas most closely associated with intelligence research—the psychometric approach, cognitive psychology, and artificial intelligence—have all shared a common neglect of motivational and contextual factors responsible for mental development and have tended to draw analogies from the type of rational deductive reasoning associated with computers. Although the deliberate suspension of motivational and contextual factors may have been a useful strategy when cognitive science was in its infancy, mental ability must now be redefined to include these neglected areas.

Social psychology is in many ways uniquely suited to address these deficiencies because motivation and the social context have been central issues in a long history of research. By "social psychology," I do not mean the narrow academic discipline typically associated with psychology but a more broad-based academic coalition consisting of psychology, anthropology, sociology,

[1] In this work, the terms *intelligence* and *mental ability* or *mental abilities* are completely interchangeable. Although the term mental abilities is more accurate because it suggests that there may be a number of different types of intelligence, the term intelligence is also used to stress that the topic being discussed is what psychologists traditionally think of as intelligence or, more specifically, the rather specialized set of skills typically associated with academic achievement.

linguistics, and history (see Collier, Minton, & Reynolds, 1991, for a more extensive discussion of this "potential" discipline). This broad-based coalition will have to link up with an equally broad coalition currently referred to as "cognitive science" (see Chapter 2), as well as such traditional areas as educational and developmental psychology. Much research has already been conducted, but researchers working in separate areas are unfamiliar with each other's findings. The contributions of these broad coalitions fall into three separate areas: (1) the role of the social context in the development of cognitive skills, (2) motivational factors that promote or inhibit the development and/or use of cognitive skills, and (3) biases and distortions in information processing that lead to errors in perception and problem solving and poor recall.

THE SOCIAL CONTEXT OF MENTAL DEVELOPMENT

At the most general level, social psychology can help show how the social and cultural context contributes to cognitive and intellectual development. A concern with the social context is not new, but it is an area that has waxed and waned. During the 1930s, there was what Boring (1950) called "an invasion of 'general psychology' by 'social psychology'" (p. 748), and a lively partnership developed between psychology and anthropology (e.g., Klineberg, 1940). Social psychologists began to examine how individual psychological processes, such as learning, perception, and memory, were shaped by social conditions. Although this interest has subsided to a large extent, there are still pockets of research that can be tapped.

The social context not only provides the information that is processed—the beliefs, attitudes, and cultural norms promoted by society—but also helps to foster the development or nondevelopment of certain types of cognitive skills. The economic and cultural context forces people to adapt by developing skills to handle new situations, which are then perfected and automated through practice. Some skills are used so habitually that they recede from awareness and become "second nature." As Vygotsky (1930–1935/1978) has pointed out:

> In psychology we often meet with processes that have already died away, that is, processes that have gone through a very long stage of historical development and have become fossilized. These fossilized forms of behavior are most easily found in the so-called automated or mechanized psychological processes which, owing to their ancient origins, are now being repeated for the millionth time and have become mechanized. (pp. 63–64)

One of the primary focuses of this book is aspects of intelligence that have become *second nature.* These are primarily cognitive skills involved in attention, perception, problem solving, and memory which have been learned incidentally for the most part, practiced to the point where they have become automatic, and are now used routinely to deal with day-to-day situations.

A second focus is on the deliberate use of cognitive strategies to aid problem solving and memory. These metacognitive skills allow people to gain control over their mental operations and improve performance. With time, these operations also become more and more automatic, so that experts within a particular domain focus primarily on the problem at hand and have very little self-conscious awareness. Both cognitive and metacognitive skills seem to develop within one domain at a time and may or may not generalize to other areas (see Ceci, 1990).

The social context leads not just to the development of skills but to maladaptive behavior as well. A good example can be seen in Carol Dweck's (1975) contrast between *mastery-oriented* and *helpless* children, which will be discussed in more detail in Chapter 3. When mastery-oriented children are faced with a temporary setback or the possibility of failure, they regard the task as a challenge, work harder, and generally improve their performance. Helpless children, in contrast, find such tasks aversive and have difficulty concentrating or sustaining interest. Their thoughts frequently wander to task-irrelevant areas, and their performance deteriorates over time. Other examples of learned maladaptive behaviour include the fear of success and failure (Chapter 3), an external locus of control (Chapter 4), and various forms of self-handicapping strategies (Chapter 5).

The significant improvement of intelligence in recent years underscores the importance of the social context. Tuddenham (1948) compared the tests of white enlisted men from World War I and World War II and found that test performance had increased by almost a full standard deviation. The median performance on the Army Alpha during World War II fell at the 83rd percentile of the corresponding World War I population. More recently, Flynn (1984) has shown that there has been the equivalent of a 13.8-point increase in IQ between 1932 and 1978. His study showed that every sample used to standardize the Stanford-Binet and Wechsler tests between 1932 and 1978 used higher norms and, thereby, made it more difficult for those who would have previously scored high to receive comparable scores. An average person scoring 100 on a test standardized in the late 1970s would have scored 114 or almost one full standard deviation above the norm on a test standardized in the early 1930s. Flynn discusses various possible reasons for the change and concludes that the most important factor is probably increases in education. Similar increases in IQ have been reported by Anderson (1982) for people in Japan.

Technology helps determine not just the "amount" of intelligence but the kind of intelligence that occurs. The advent of low-cost home computers, for example, has helped produce a new generation of computer whiz kids in which programming skills and certain types of logical thinking are more or less second nature. Commercial television, on the other hand, seems to limit people's ability to attend to information and to concentrate for extended periods (see Chapter 7). Although there has been a common tendency to see intellectual development as something that *evolves,* modern industrial conditions both promote and limit cognitive development and there are enormous variations between individuals and subgroups even in the same culture. If specific skills can be altered so radically in such a short period, there is no reason to assume that even more radical differences across cultures and subcultures have not produced similar effects. A person who is quite able to adapt to society at one time may feel extremely out of place once cultural conditions change. Indeed, many middle-aged white-collar and blue-collar workers in our society are in very real danger of being left behind by the computer revolution.

Research on cross-cultural differences in intelligence has frequently been avoided by more progressive thinkers because there is a common tendency to see intelligence along a continuum that ranges from "primitive" to "civilized," with the implication that civilized is better (e.g., Werner, 1948). The basic premise behind what Medin and Cole (1975) have called the "search for historical antecedents" is that subjects who are younger or are from less technologically advanced cultures share certain characteristics that vary along a scale from primitive to advanced. The ethnocentrism and chauvinism of this position is sufficiently strong to dissuade many progressive people from even attempting research in the area. What is needed is a radical realignment so that a concern with cultural differences is viewed as progressive. Culture helps shape intelligence, for better or worse, because it determines the kinds of opportunity available (a person cannot become computer literate in a society without computers). Individual differences occur, at least in part, because of individual differences in cultural conditions and, therefore, improvements among disadvantaged groups can be expected only when resources are more widely shared.

There is *no* strictly linear continuum which runs from "primitive" to "advanced." The cultural context occurring in modern industrial society facilitates the development of some skills and limits others. Cole and Means (1981) point out that "departures from typical performance patterns in American adults are not necessarily defects, but may indeed be excellent adaptations to the life circumstances of the people involved" (pp. 161–162). Similar differences occur across subcultures even in the same society. What is adaptive for a young black slum dweller may be very different from what is adaptive for a wealthy middle-aged professional. Atkinson (1974) has stressed that scores on *so-called* intelligence tests measure performance on particular tasks in a particular situation. He goes on to suggest that "one really needs a whole social psychology to make complete coherent sense out of what happens in that setting. *Mental testing is, or should be, a subfield of social psychology"* (p. 406). The social environment also contributes indirectly to mental development through the development of various motives that then help determine what a person learns or fails to learn.

MOTIVATION AND MENTAL ABILITY

Social psychologists have long been involved in research on motivational factors directly or indirectly involved in mental ability. This includes research on achievement motivation, internal-external locus of control, level of aspiration, intrinsic reinforcement, and most recently, various forms of self-handicapping strategies. Many of these areas have a long history of research. Research on achievement motivation can be traced back to Max Weber's (1904–1905/1958) description of the Protestant work ethic, discussions of intrinsic reinforcement draw partly from Gordon Allport's (1937) theory of "functional autonomy," Leon Festinger's doctoral dissertation, completed under Kurt Lewin in 1938 was based on level of aspiration, and Julian Rotter's (1954, 1966) research has helped inspire a swarm of studies on attribution and locus of control. The most recent work on self-handicapping shows that people do not always aspire to succeed. They frequently create obstacles so that they can attribute failure to sources other than their own ability.

Each of these motives has different social origins, different cognitive consequences, and a somewhat different effect on various target groups. The early research on achievement motivation described in Chapter 3, for example, was carried out primarily with males, and there was an early recognition that the same processes were not consistently related to achievement motivation in females. Because of this, Martina Horner (1968) postulated an additional motive—the fear of success—to help explain lack of achievement among certain females. The more recent research based on attribution theory has also found consistent gender differences in the way males and females explain success and failure. In general, males take more credit for success than females and are more likely to deny responsibility for failure. This helps perpetuate the myth of male superiority and discourages high levels of achievement among females.

Although gender differences also occur in locus of control, the most visible differences are those occurring between blacks and whites. Among white males, an internal locus of control in which people believe that they can control what happens to them is usually associated with a warm and protective home environment that stresses independence and frequently rewards success. Among blacks, an external locus of control seems to be based on black children's growing awareness that lower-class blacks have less control over their own lives. An adequate understanding of these processes requires an understanding of the conditions of blacks in the United States today.

Similar differences among larger groups are also evident in the literature on level of aspiration and self-esteem discussed in Chapter 5 and in the research on intrinsic motivation, which is the focus of Chapter 6. Level of aspiration is typically lower among underclass blacks, and intrinsic motivation is rare among people in jobs that have been oversimplified and reduced to a few simple repetitive operations. These effects are often subtle and indirect. Intrinsic interest in a particular activity, for example, helps to determine the amount of information that is learned, which in turn affects people's ability to acquire more information, recall previous information, and use the information to solve new problems.

Motivation affects mental ability in three ways. First, certain forms of motivation, such as intrinsic motivation, achievement motivation, an internal locus of control, and high levels of aspiration, promote the development and use of cognitive skills. People with high achievement motivation select tasks that are moderately difficult and challenging and are more persistent and goal oriented, as well. Those with an internal locus of control have been found to seek out information more often and process, recall, and use it more effectively. Intrinsic motivation and high levels of aspiration promote the development of academic and nonacademic skills, leading ultimately to high levels of knowledge that make information processing more rapid, more structured, and more profound.

Second, motivation, as mentioned previously, can also inhibit the development of cognitive skills or lead to the development of processes which are counterproductive. Helpless children or those with a high fear of failure, for example, have difficulty concentrating when tasks become difficult. They are more likely to devalue the tasks and give up. The need to maintain high levels of self-esteem causes people to place obstacles in their path so that they will have an excuse for their poor performance. The most common obstacle is simply not to try at all. For a variety of reasons, the correlation between measures of motivation and academic achievement are typically quite small, but there is an additive effect in that those who experience low achievement motivation often experience low levels of aspiration, intrinsic motivation, and self-esteem plus an external locus of control.

Third, there is a complex interplay between motivation and metacognitive skills that is just now beginning to be explored (see Pressley, Borkowski, & Sullivan, 1985; Schneider & Weinert, 1990). Intensive training in various memory strategies, for example, leads to the recognition that there is a direct link between effort and good recall, and this metacognitive awareness leads to increased effort in the face of potential failure. It is thus possible to undermine the "helpless" pattern described by Dweck (1975) by providing people with explicit knowledge of their cognitive skills. Motivational factors determine not just the goals toward which people aspire but the way in which they seek out, process, and use information.

BIASES IN INFORMATION PROCESSING

Social psychology can also contribute to our understanding of mental ability through a more precise understanding of the distortions and biases that occur during information processing. Many social psychologists, like their cognitive counterparts, began with the assumption that people process information more or less rationally, using all the information currently available (e.g., Jones & Davis, 1965; Kelley, 1967, 1972). It soon became apparent, however, that observers do not use all the available information (e.g., McAuthur, 1972). Judgments are often made quickly with limited data that are haphazardly combined and strongly influenced by preconceptions. People overestimate their own contribution to shared activities or the extent that others agree with them. They give themselves more credit for success and take less credit for failure than they deserve. Some of these distortions occur because of motivational factors, but others are based on information processing itself (Miller & Ross, 1975).

Biases and distortions in information processing occur for two basic reasons. First, people's beliefs often distort the way they process new information. Errors in person perception are more apparent than errors about objects because the labels used to describe people—"masculine," "feminine," "rich," "poor," "intelligent" or "stupid"—have fuzzy boundaries and mean different things to different people. A considerable body of research has shown that the labels people use to describe other individuals affect the perception and recall of information. People are more likely to process information that is consistent with the labels. The prejudices and preconceptions that people bring to a situation help determine how they perceive themselves, other people, and the world.

People make mistakes not simply because they lack information but because they do not use all the information that they do have. They tend to be "cognitively lazy" and process information only until they are sufficiently certain that their judgments are correct. They commonly use shortcuts and heuristics that are right most of the time but can be wrong. The biases are not random, however. They can often be explained by clearly defined laws, and the study of these laws is currently at the center of experimental social psychology.

The kinds of error that people make in person perception may occur in information processing in general. People make systematic errors in problem solving, perception, and recall depending on their background and previous experience. The distinction between people who are "reflective" and "impulsive," for example, shows that impulsive people do not take sufficient time to process information and jump to premature conclusions which are often wrong (Baron, Badgio, & Gaskins, 1986; Kagan, 1971). Others who have been raised in authoritarian homes become dogmatically fixated on conventional solutions to complex problems and fail to consider other alternatives. The quality of creative problem solving is determined by the amount of time and effort people are willing to spend studying problems or verifying the solutions. Work on problem solving and memory suggests that people often fail to use even well-learned strategies and, therefore, perform poorly or forget details. This research shows that there is a distinction between how people perform (e.g., on academic tasks or IQ tests) and what they can or will do, given sufficient motivation and interest. The failure to use skills, on the other hand, affects not just the accuracy of the current response, but overall level of ability, because frequently used skills quickly become habitual and are carried out with very little effort.

VYGOTSKY'S ZONE OF PROXIMAL DEVELOPMENT

Much of the recent research on the social origins of mental ability has been guided or inspired by the work of the Russian psychologist, Lev Vygotsky (1896–1934), who worked and wrote early in the twentieth century. Vygotsky was trained in literary criticism and did his doctoral dissertation on Shakespeare's *Hamlet,* but his interests began to shift to language and eventually to the psychology of language, thought, and mental development. Vygotsky felt that an adequate understanding of the social origins of mental ability must begin with the concept of *practical activity.* Children develop intellectual skills by adapting to their environment; and more basic processes, such as perception and memory, are transformed in the process. Vygotsky (1930–1935/1978), for example, recognized two forms of memory—natural and symbolic—and felt that the essence of human memory was based on an ability to actively describe the original events. All higher order skills are mediated processes based on social interaction. For Vygotsky, every higher function appears twice: first on the social level between people and then on the individual level within people.

One of Vygotsky's (1930–1935/1978) central concepts is his "zone of proximal development," which he defines as "*the distance between the actual developmental level as determined by independent problem solving and the level of potential development as determined through problem solving under adult guidance or in collaboration with more capable peers*" (p. 86, italics in original). The zone of proximal development refers to the difference between the level of performance children can reach unaided and the level that can be accomplished when guided by someone with more knowledge or experience. It is jointly determined by the form of instruction and the current level of mental ability.

Vygotsky's concept of a zone of proximal development has inspired contemporary researchers in a number of different ways. First, it provides a *different conception of intelligence.* Instead of seeing intelligence as a static entity based primarily on genetic endowment, Vygotsky stressed the difference between what children are capable of doing on their own and what they can do given proper guidance or instruction. He rejected the notion that intelligence can be measured by a single test, such as the IQ test, and pointed out that two children with similar IQs may differ greatly in their capacity to learn.

To illustrate, Vygotsky (1930–1935/1978) used the following hypothetical example: Suppose we have two 10-year-old children with a mental age of 8—that is, they can complete tasks typical of those who are 8 years old without assistance. With guidance and instruction, the first child learns to complete problems typical of 12-year-olds, whereas the second can function only at the level of a 9-year-old. He then asked (somewhat rhetorically), "Are these children mentally the same?" (p. 86) and concluded that, in an important sense, they are not.

The difference between 8 and 12 in the first case and 8 and 9 in the second is the difference between the "actual developmental level" and the level of potential development, which forms the zone of proximal development. The actual developmental level is measured by traditional IQ tests, which serve as a summary of what children can do independently or what they already know. The level of performance after instruction helps us identify those functions that have not fully matured but are currently in the process of maturing. Vygotsky likened them to the "buds" or "flowers" of development rather than the "fruits" of development. Identifying the zone of proximal development allows us to determine children's immediate future and their dynamic development. He points out that "what is in the zone of proximal development today will be the actual developmental level tomorrow" (Vygotsky, 1930–1935/1978, p. 86).

A second line of research that is central for both Vygotsky and this book focuses on the *development of cognitive skills* during social interaction. For Vygotsky, higher mental skills are qualitatively different from elementary processes and can be distinguished by four somewhat overlapping criteria (see Wertsch, 1985): (1) Elementary processes are primarily under what is now called stimulus control, whereas higher processes become more and more voluntary; (2) higher processes can be consciously regulated; (3) higher mental skills are mediated by language; (4) higher processes are acquired through social interaction with others who are more skillful or knowledgeable. Supervision and guidance allow less experienced people to participate in and complete tasks that would have been impossible alone. Elementary processes, in contrast, have biological roots, are primarily under the control of the environment, and are not consciously regulated or mediated by language. The primary focus of this book is the development of these higher mental skills.

A third contribution, which is somewhat tangential to the theory as a whole, is Vygotsky's (1933/1966) treatment of *play* and its role in mental development. Whereas the previous description of his zone of proximal development stresses the importance of instruction and social interaction, his treatment of play makes it clear that it also plays a role in self-motivating, intrinsically interesting activities. Play is a "leading activity" or central goal among preschoolers, in which children experiment with meaning, master new skills, and develop a detached perspective. His notion of play is important for the present discussion because it links his concept of cognitive development to motivation. Although this aspect of his theory was not well developed, Vygotsky (1933/1966) pointed out:

> The mistake of a large number of accepted theories is their disregard for the child's needs—taken in the broadest sense, from inclinations to interests, as needs of an intellectual nature—or, more briefly, the disregard of everything that can come under the category of incentives and motives for action. We often describe a child's development as the development of his intellectual functions, i.e., every child stands before us as a theoretical being who, according to the higher or lower level of his intellectual development, moves from one stage to another.
>
> Without a consideration of the child's needs, inclinations, incentives and motives to act—as research has demonstrated—there will never be any advance from one stage to the next. (p. 7)

According to Vygotsky, play creates its own zone of proximal development—"In play it is as though he were a head taller than himself" (p. 16).

Finally, Vygotsky's theory can serve as a useful model for *educational reform.* Vygotsky (1930–1935/1978) believed that instruction is useful only when it precedes development and awakens processes that are within the zone of proximal development. Children first observe while adults guide their activities, but gradually begin to participate and take more and more responsibility. By communicating and describing what is going on, children not only develop skills but gain the knowledge necessary to use those skills at the right time and consciously regulate their own behavior.

A good example of the direct application of Vygotsky's theory to education is Palincsar and Brown's (1984) research on *reading comprehension.* These researchers begin by identifying four activities—summarizing, questioning, clarifying, and predicting—that have been found in almost every study of critical reading and used a method of "reciprocal teaching" to teach these skills to seventh-grade students who were two to three years behind in their reading ability. They point out that such skills serve a double function of increasing comprehension and providing feedback about how much is understood. If readers cannot provide an adequate synopsis, make up questions, clarify difficulties, or predict what may occur, then this is a good sign that comprehension is not taking place.

In the method of *reciprocal teaching,* a teacher and a student take turns reading a passage and commenting on it. At first, poor readers had difficulty participating in the dialogue, asking questions, and summarizing material. The adult teacher provided a variety of prompts, which included making up questions and paraphrasing statements for the student to repeat verbatim. After about 10 sessions, students became much more sophisticated in their ability to ask questions and describe the material in their own words. By the end of the sessions, these procedures had brought students from the very bottom to the upper half of their class. Although some loss occurred over time, these losses were quickly made up by a single refresher session. What is more, trained students developed the capacity to serve as peer tutors and go on to help others with similar problems. The method of reciprocal teaching can thus supplement or replace traditional approaches based on lectures. If Vygotsky's theory is correct, it may be necessary to more closely tailor the instruction to children's current level of ability and peer tutoring serves as a reliable and inexpensive way of achieving this goal. These and similar developments will be discussed more fully in the final chapter of this book.

Although Vygotsky's theory serves as a useful starting point for organizing much of the literature on cognitive development, the current approach draws from a number of other theoretical perspectives. There are several reasons for this. First, as mentioned previously, much of the research on motivation and cognition has been inspired by theories quite unlike Vygotsky's. These theories have had a long history of research by people who have made little or no effort to link them with Vygotsky. His theory, although quite old, has only recently begun to have a major effect on psychological research in America, and although a summary of this research would be useful, it is not the focus of this book.

A second reason for not limiting the current review to Vygotsky's theory is that Vygotsky's approach has several shortcomings and weaknesses (see Wertsch, 1985). First, Vygotsky focused on the social origins of mental ability and did not deal with biological aspects of human development to any great extent. The present focus is also on the social origins of mental ability, but certain aspects of cognitive development, such as perception and language development, must be seen within the context of biological maturation. Recent research on perceptual development, for example, has shown that newborns come into the world with an extensive array of innate skills and an incredible capacity to learn from certain types of experience. Second, although Vygotsky recognized the importance of motivation in cognitive development, this too was a relatively neglected area. Since his untimely death at the age of 37, an extensive body of research has been developed on the role of motivation in cognitive development and this literature will be reviewed in Chapters 3 through 6. Finally, although Vygotsky's attempt to reformulate psychology along Marxist lines called for an account of the relationship between the conscious and the larger social context, his own work focused almost exclusively on face-to-face interactions. As Wertsch (1985) has pointed out, because Vygotsky did not consider the larger cultural context, he did not provide a comprehensive account of the social origins of mental ability, even though his position is compatible with such an account.

For all these reasons, Vygotsky's theory needs to be supplemented by more broad-based social theories, as well as a concrete understanding of social conditions within contemporary society. The current approach focuses on specific motives and cognitive skills learned during the course of social interaction and automated through practice. These can be seen as part of an intermediate step in a causal chain that begins with social conditions and ends with performance on tests designed to measure differences in mental ability. Those trained in the experimental tradition may find this approach a bit disconcerting, since it blends "hard" experimental data with "soft" data derived through other procedures, but it is absolutely essential for understanding the social origins of mental ability.

The current approach stresses *differences rather than deficiencies* in mental ability, but this point is easily lost because the main focus is on skills occurring in academic settings. The nature–nurture controversy is particularly strong in this area, and an adequate understanding of the role of the social environment may help clarify the problem. An example of a difference model that begins to look like a deficiency model can be seen in Brown, Palincsar, and Purcell's (1986) research on early reading ability. These researchers point out that middle-class mothers often begin to read to their children very early, sometimes before they are one year of age. Middle-class mothers point out pictures in books and describe them in terms of essential features. As these children grow up, they are gradually given more and more responsibility for identifying and describing objects in books. These "known-answer" questions help prepare middle-class children for school because they are precisely the same type that teachers ask in the classroom.

Lower-class black children, on the other hand, have less access to reading material. They are often asked to describe objects in terms of similarities rather than features and respond to questions such as "What is this *like?*" They, therefore, develop a complex form of metaphoric thinking that they use in such games as "playing the dozen." In this game, young boys try to outdo each other by saying such things as "Your mother is like . . . " This is usually seen as a form of good-natured teasing, and it can lead to some truly creative responses. But these skills are rarely used in the classroom. Heath (1989) argues that many poor blacks are brought up to believe that those in authority do not reward obvious answers but look for more imaginative or creative responses. When poor black children are asked a question that the teacher, who is after all a competent adult, obviously can answer, such as "What is this—a cat or a dog?" the question seems strange and unfamiliar. Poor black children often respond by not responding or by looking surprised and confused. The teacher takes this lack of response as a lack of knowledge and often assumes that children who cannot answer such obvious questions must be stupid or even mentally retarded. They then stop asking questions so as not to embarrass youngsters who do not respond properly, thus inadvertently leaving them out of the education process. Such children then fall further and further behind.

The preschool experience of middle-class children, both black and white, is to a large extent a *preparation* for school, since it involves tasks and interactions similar to those within the school. Poor black children, on the other hand, have a different set of experiences and develop a different set of skills, which are extremely useful in terms of creative problem solving but have no place in the classroom. The inability of white middle-class teachers to recognize these skills is due to an easily corrected ignorance of black society—what Jacqueline Irvine (1990) calls a lack of "cultural sync." What is equally disturbing is that a very useful form of metaphoric thinking is not developed, and it is often seen as a form of disruptive behavior.

The central assumption of this work is that people acquire cognitive skills by adapting to their environment and interacting with other people. The skills most frequently used improve with practice; whereas those not needed may atrophy or fail to occur. Different environments elicit different skills so that each person develops a repertoire of abilities and disabilities that is somewhat unique. By the end of the book, it should be apparent that social factors play a major role in both motivation and cognition and may well account for existing differences in performance between groups.

Mental ability can be seen as a combination of genetic potential and acquired motivation and cognitive skills that allows people to deal with the situations and solve the problems they commonly confront in their environment. Because the focus of this book is on the *social* origins of mental ability, discussion of the genetic contribution will be minimal. The discussion will concentrate on skills involved in academic performance and traditional intelligence tests. Although more comprehensive models of human intelligence have been developed (e.g., Gardner, 1983), I will not address such topics as musical ability, physical coordination, or social sensitivity, even though these are legitimate areas of concern. Motivation will be covered in Part Two and this will be followed by a discussion of acquired cognitive skills. Implications for education and social reform will be discussed in the final chapter.

CHAPTER 2

Current Controversies and Converging Trends

Several controversial aspects of intelligence research are examined, including the concept and measurement of intelligence and the relative contributions of nature and nurture. The impact of the cognitive revolution, with its heavy emphasis on computer models, is reviewed and it is argued that, although computer models have been useful in generating research, fundamental differences in the way people and computers process information must be taken into account. Finally, the "new look" in intelligence research inspired by the cognitive revolution is briefly considered. Although the more recent accounts of intelligence represent a considerable advance over previous psychometric approaches, several aspects of intelligence, including the role of motivation and the larger social context, are still relatively neglected.

Most people associate intelligence with a number of discrete skills such as the ability to communicate, perform well academically, think quickly, make money, play chess, or create works of art. Some people are street smart but do poorly in school, others are market wise but have few social graces and little interest in art. Some people with highly refined technical skills show an appalling ignorance of social problems and a remarkable insensitivity to those most severely affected.

Most people are also aware that there is a considerable psychological industry devoted to the assessment of mental ability. Few people in our society have escaped psychological testing. These tests are given routinely to identify children with problems requiring remedial attention or to determine placement in school. Achievement tests are given near the end of high school to select students for university and at the end of university to determine entrance into professional programs and graduate schools. Professionals in areas such as law, medicine, and clinical psychology take comprehensive examinations at the end of their training that determine whether they can or cannot work in their chosen fields. Psychological testing has become a multimillion-dollar industry and a fact of life for most people in our culture. So it may come as a bit of a surprise to discover that there is still considerable controversy over basic questions and a number of unresolved issues. The three most central are:

1. What is intelligence?
2. Can it be measured?
3. What is the relative contribution of heredity and the environment?

The first two questions are very much related. The inability to define intelligence makes it very difficult to measure. There is no commonly accepted definition of intelligence and, therefore, no accepted measure.

ASSESSMENTS OF MENTAL ABILITY

Intelligence tests, as we know them, can be traced to the work of the French psychologist Alfred Binet (1857–1911), at the turn of the century. Binet was commissioned by the Ministry of Education in France to provide a test that could identify children unable to perform well in school. Binet and his colleague, Théophile Simon, assembled a sample of questions and puzzles related to academic performance and published the Binet–Simon Scale in 1905. Each question was assigned an age level based on the typical performance for that age group. A "successful" item was one that 60% of the children of a particular age could complete and that significantly fewer children one year younger could answer correctly. Examples of test items from the 1908 Binet–Simon Scale include the following:

Age	*Item*
3	Point to the nose, eyes, and mouth.
4	Repeat three digits.
5	Choose the heavier of two cubes.
6	Define familiar objects in terms of use.
7	Tell what is missing in an unfamiliar picture.
8	Count backward from 20 to 0.
9	Arrange five blocks in order of weight.
10	Construct a sentence using three words—*Paris, fortune, gutter.*
11	Point out absurdities in contradictory statements.
12	Repeat seven digits.
13	Give differences between pairs of abstract terms such as *pride* and *pretension.*

Children began with the easiest items and completed as many as possible. Their level of performance, or *mental age,* was based on the last (or most difficult) set of items completed. The measure of intelligence was obtained by subtracting the mental age from the chronological age. In 1912, the German psychologist William Stern suggested that the mental age should be divided by the chronological age to produce a relative measure, and the concept of *intelligence quotient,* or IQ, was formed.

Binet's test was a purely diagnostic instrument designed to identify slow learners so that they could receive remedial attention, but when the test was imported to the United States, it was widely regarded as a measure of a person's actual ability. The person most responsible for promoting the intelligence test in the United States was Lewis M. Terman of Stanford University (see Minton, 1988). He translated the Binet–Simon Scale, increased the number of items, and created the Stanford–Binet (Terman, 1916). His scale contained a revised formula for computing IQ. The intelligence quotient was obtained by dividing the mental age, based on test performance, by the chronological age and multiplying by 100. Those with average scores have the same mental and chronological age and score 100 on the IQ test. Those above and below average have scores above and below 100. Group intelligence tests were widely used to screen recruits during World War I (Yerkes, 1921), and although their success has been greatly exaggerated, testing helped establish psychology as an applied science (see Gould, 1981).

Although intelligence is often conceptualized as a unitary phenomenon, there has been considerable disagreement among those most directly involved in psychological testing. Even investigators who view intelligence as a global trait acknowledge that people have specific abilities that are somewhat independent. Charles Spearman (1904, 1927), for example, developed and used factor analysis to explore intelligence and concluded that it was made up of two components—a general factor, *g,* and specific factors, *s.* A person's performance on a particular task is based on both his or her general ability and specific ability in a particular area (such as verbal fluency, spatial ability, or mathematics).

Some researchers, such as Louis Thurstone, have questioned the concept of "general" intelligence altogether. Thurstone (1938) applied factor analysis to a battery of 56 tests and identified seven "primary mental abilities": *verbal comprehension, word fluency, number facility, spatial visualization, associative memory, reasoning,* and *perceptual speed.* The most complex model of intelligence is probably that suggested by Guilford (1967), who initially developed a theoretical model with 120 factors. His model has recently been increased to 150 factors located along three dimensions (Guilford, 1977, 1985). His first dimension is a content dimension that describes the type of information processed and consists of visual, auditory, symbolic, semantic, and behavioral components. The second dimension consists of operations, such as evaluation, convergent production, divergent production, memory, and cognition. The third is a product dimension that specifies the form the information takes and includes units, classes, relationships, systems, transformations, and implications. These generate 150 possible "varieties of intelligence," including such diverse skills as social intelligence and discrimination of melodies and rhythms. Although Guilford does not claim that each is completely independent, he estimates that he and his colleagues have identified about a hundred of these to his satisfaction (Guilford, 1982). Although Guilford's model has been criticized for being excessively complex, he argues that intelligence is a complex concept that requires comprehensive models to do it justice.[1]

More recently, Gardner (1983) has attempted to divide intelligence into categories based on special skills and localized regions within the brain. He has identified six types of intelligence—linguistic, musical, logical-mathematical, spatial, body-kinesthetic, and personal. Body-kinesthetic intelligence includes the skills involved in sports, typing, and dancing, whereas personal intelligence is based on a person's sensitivity to his or her own feelings and those of others. In our culture, it would not occur to many people that a person who is unmusical, uncoordinated, or lacks social skills is unintelligent but this is only because we have learned to equate intelligence with academic success. If we take a more biological view, we may have to concede that intelligence comes in many different forms.

Thus, after more than a century of research, there is still considerable confusion about the nature of intelligence and our current ability to measure it. Part of the problem is based on definition. After all, "intelligence" is merely a concept (see Sternberg & Ditterman, 1986). It can be narrowly defined in operational terms as "the capacity to do well on an intelligence test" (Boring, 1923, p. 35), or it can be defined more broadly so as to include a wide range of skills not measured by any existing test. Neisser (1978) has argued that the concept of intelligence cannot be explicitly defined because it includes a number of nonoverlapping traits connected with each other through a prototype or "family resemblance." He points out that

[1] Some of the discrepancies between various models of intelligence occur because the dimensions derived through *factor analysis* depend on the tests developed to measure intelligence and ultimately on the individual items. Factor analysis is a correlational procedure for identifying *clusters* of relationships among a large number of items. Those highly correlated tend to be identified as a common factor. Each test contains a sample of items supposedly related to intelligence, but there is no reason to assume that it samples all possible cognitive skills. The skills most frequently tested are those related to academic performance, so it is not surprising that there is a moderately strong correlation (from .40 to .60) between IQ and success in school.

The construction of IQ tests, such as the Wechsler Adult Intelligence Scale or the Stanford–Binet, rest heavily on the concept of general intelligence. Items and subscales that fail to correlate sufficiently with the test as a whole are simply eliminated and only those with moderately high correlations are retained. This gives the impression that all items measure the same thing. Factor analysis does not give unequivocal results even when the same items are used (Lewontin, 1970). One strategy in factor analysis is to pack as much as possible into a single factor, whereas another is to distribute the weights over as many factors as possible. This explains why two investigators can come up with very different interpretations for the same data.

"two people may both be quite intelligent and yet have very few traits in common—they resemble the prototype along different dimensions."[2]

NATURE VERSUS NURTURE

A second controversy that has plagued psychological testing is the relative contribution of heredity and the environment. The nature–nurture debate about intelligence is one of the oldest and most heated controversies in psychology. Interest in this area stems largely from the evolutionary theories of Charles Darwin (1809–1882) and Herbert Spencer (1820–1903) and the more systematic work of Darwin's cousin, Sir Francis Galton (1822–1911). Galton was involved in a number of scientific areas, but he is best known for his work on individual differences and the application of statistical procedures to the study of heredity. Galton was struck by the observation that *eminence* seemed to run in families, including his own. He examined the family trees of different men of eminence and found that the proportion of eminent men with eminent relatives was higher than that of eminent men in the general population. The notion that intelligence is inherited was so original that it surprised even Darwin, who attributed individual differences to the degree of effort.

Galton's belief in inherited mental ability led to the idea that the human race could be improved through selective breeding—a position that he labeled "eugenics." Galton believed that social improvement could only come about by identifying individuals with superior traits and eliminating defective ones from the gene pool. His subsequent attempts to measure mental ability were designed to put eugenics on a scientific footing. While he recognized the role of the environment, he stressed genetic factors and felt that any person with sufficient intelligence and a desire to succeed could overcome environmental obstacles and become a success. This assumption overlooked and in fact justified the great advantages enjoyed by the British upper class—including Galton himself.

By the time Galton died in 1911, the eugenics movement had become quite strong in both Britain and the United States. A Eugenics Record Office was set up on Long Island, with financial support from John D. Rockefeller and others, to determine "America's most effective bloodlines" (Gould, 1981). Courses in eugenics were being given at major colleges across the country, and eugenics was widely regarded as a biological panacea for all social problems. Intelligence tests were later used to rank various racial and ethnic groups. H. H. Goddard (1917) gave intelligence tests to immigrant groups at Ellis Island in New York and concluded that 83% of Jews, 80% of Hungarians,

[2] There are also large cross-cultural differences in the concept of "intelligence." African societies, for example, typically place less emphasis on speed and more emphasis on social skill.

79% of Italians and 87% of Russians were feeble-minded. In 1924, new quotas were established that virtually eliminated immigration from southern and eastern Europe.

The British wing of the eugenics movement was kept alive by such notable figures as Karl Pearson (1867–1936), who developed the correlation coefficient, and Sir Cyril Burt (1883–1972). Burt was a disciple of Galton, a successor of Spearman, and the first psychologist to have been knighted for his work. His early research used teachers' ratings of "intelligence" as a validating criterion and generally confirmed Spearman's conclusions about general and specific abilities. Burt went further, however. He compared upper-class boys in private schools to lower-class boys in local grammar schools and concluded that the upper-class boys were inherently superior (Burt, 1966). Burt's best known work was based on the study of twins and people with varying degrees of genetic relatedness. Although Burt was not the only person working in this area, he was by far the leading figure. His reputation rested on several factors—his statistical expertise, the apparent size of his sample, and his inclusion of a number of more remote kinship relations, such as uncle–nephew and second cousins (Hearnshaw, 1979). His work and others comparing people with various degrees of genetic relatedness led Jensen (1969) to claim that heredity accounts for about 80% of the variance in IQ scores.[3]

The genetic position was seriously weakened when Burt was discovered to have fabricated much of his data. Although he carried out some studies comparing identical and fraternal twins before World War II, he continued to publish after the war without new data. Leon Kamin (1972, 1974), an American critic of the genetic position, examined Burt's later work and found that although the number of reported cases of twins continued to increase, many of the correlations remained precisely the same. Later Leslie Hearnshaw (1979), an American psychologist who was asked to write Burt's biography, found that no new subjects had been tested during the later part of Burt's life and that Burt had probably published work using the names of several fictitious assistants. Burt was apparently responding to several waves of criticism and wanted to create the impression that, unlike his critics, he was actively involved in ongoing research.

A reassessment of twin studies without Burt's data lowers the estimate of the genetic contribution to about 60% (Wade, 1976). Others have argued,

[3] The concept of *heritability* is somewhat confusing. It does not describe the relative contribution of genetic factors. It reflects the proportion of *variance* accounted for by genetics (that is, the variance of heredity/total variance). Ceci (1990) points out that environmental factors can play a major role even when heritability estimates are extremely high. The heritability of height, for example, is near 1.0, and yet Japanese raised in the United States are more than 5 inches taller than those living in Japan (Grevlich, 1957). American and British teenagers are also 6 inches taller than they were a century ago. If there is little variance, on the other hand, then heritability will be low. Because there is little variation in the number of ears people have, the heritability of having two ears is approximately zero despite the obvious role of genetics.

however, that separated twins do not provide a definitive test because they are similar in other ways and are raised in similar environments. In many cases, separated twins lived in the same town, attended the same schools, and were raised by relatives (Kamin, 1974). Vernon (1969) has suggested that, if separated twins were raised in radically different cultures, the relative contribution of genetics would be much lower and possibly even reversed.[4] Several reviews (Henderson, 1982; Plomin, DeFries, & McClearn, 1980) have pointed out that the heritability estimates for general intelligence vary with the date of publication. Older studies published prior to 1963 provided the bases for the often cited 80%, whereas those published after 1975 seem to have converged on a figure closer to 50%.

An accurate assessment of the relative contribution of nature and nurture is made more difficult because *covariation* between the two are often treated as genetic. Gifted children, for example, may increase their intelligence by seeking intellectual stimulation, through reading, puzzles, and so on. Such children may be more interesting to talk with and receive more attention from parents and teachers. Small differences in intelligence among preschool children may be magnified once a child enters school. Sources of variance that are not explicitly due to the environment are often grouped together under the heading "covariance" and are counted as part of the genetic component, even though they are due to both environmental and genetic factors (Block & Dworkin, 1974).

The nature–nurture debate is fueled by differences between racial groups. This conflict too has a long history. Galton felt that different environments placed different pressures on various groups and led to the genetic inheritance of different traits. The harsh northern conditions and bellicose nature of the "white" race had eliminated many of the less intelligent, whereas the milder conditions in Africa had not. The result was a large number of "half-witted men" among the black population. Today, it is widely known that black Americans score an average of 15 points (or about one standard deviation) below white Americans on standard IQ tests and these differences are often attributed to differences in genetic potential (e.g., Jensen, 1969).

These two issues should be kept separate, however. Support for genetic factors within relatively homogeneous populations, such as white twins, in no way supports the claim that ethnic and racial groups differ genetically. The cultural and socioeconomic conditions of blacks in American society are so fundamentally different from the conditions of whites that, without a better

[4] Bronfenbrenner (1975) has recalculated correlations for pairs of separated twins and has found large differences for similar and dissimilar environments. The correlation between IQ scores for those raised in the same town was .83, but for those raised in different towns, the correlation was .67. The correlation for those attending the same school was .87, but this dropped to .66 for those attending schools in different towns. When communities were classified as similar or dissimilar on the basis of size and economic base (e.g., mining versus agriculture), the correlation for separated twins living in similar communities was .86 but it was only .26 when towns were not alike.

understanding of the role of social factors in intelligence, the two groups cannot be compared (see Helms, 1992, for a more extensive discussion). When black children have been adopted and raised in upper-middle-class white homes, their average IQ has exceeded the white norm (Scarr & Weinberg, 1976). Flynn (1987) points out that this comparison may actually underestimate the role of the environment, because adopted black children probably experience racism outside the home. Scarr, Pakstis, Katz, and Barker (1977) used blood tests to examine the connection between degree of black ancestry and IQ scores, but found no relationship. Motivation may also play a role. Johnson, Bradley-Johnson, McCarthy, and Jamie (1984), for example, found that inner-city black children scored an average of 13 points higher on IQ tests when they were given tokens as a reward for correct answers. Tokens had no effect of the performance of white middle-class children, presumably because they were intrinsically motivated to begin with.

It is not the intention of this book to debate the relative contribution of nature and nurture (for more extensive reviews, see Ceci, 1990; Mackenzie, 1984). More than 30 years ago, Anne Anastasi (1958) suggested that, instead of comparing the relative contribution of nature and nurture or debating which is more important, we should be concentrating on how nature and nurture interact. It is possible to concede that there is a *significant* genetic component associated with intelligence and still allow room for considerable variation based on socialization and culture. Bright parents tend to have bright children and this is due to both genetics and the environment.

It may well turn out that differences in intelligence are linked to clearly defined neurological factors, such as the extent of myelination, the relative proportion of neural transmitters or inhibitors, or the complexity of the neural network and the number of synaptic junctions. Specific skills may be based on the relative size of brain areas devoted to a particular task (Gardner, 1983). There is little reason to assume that physiological differences associated with intelligence vary with race, however, and there is also considerable evidence that the size and the structure of the brain itself are affected by the environment.

THE COGNITIVE REVOLUTION

Interest in mental ability has not been limited to researchers in psychometrics. Educational psychologists, for obvious reasons, have been very involved in assessing and improving performance in specific areas. Changes in mental ability as a result of maturation and experience have been central topics within developmental psychology. Linguists and anthropologists have been concerned with cross-cultural differences. Philosophers are engaged in a lively debate about the nature of intelligence, and the answers to these and related questions may have a major impact on political and economic policies. The field of intelligence testing has also been recently transformed by the introduction of a relatively new discipline—*cognitive science,* which includes a blend of cognitive psychology, philosophy, linguistics, and neuropsychology (see Gardner, 1985; Stillings, et al., 1987 for reviews).

Cognitive science draws heavily from the field of *artificial intelligence* or the use of computers to simulate human intelligence. The idea behind the modern computer can be traced to the British mathematician, Alan Turing (1937), who developed the notion of a simple computing machine capable of carrying out mathematical operations. A Turing machine, as it became known, is simply a scanner that can read a tape of instructions printed in binary code, similar to the on–off function of an electric light switch. Turing proved mathematically that such a machine could, at least in principle, carry out any logical task, provided the tape was sufficiently long and properly coded. In the 1940s, John von Neumann developed the idea of stored programs, that provide a set of instructions within the computer to direct and control its operations. Almost immediately, certain similarities between computers and the human brain were noticed. McCullough and Pitts (1943) showed that the operations of nerve cells could be formulated in logical terms that resemble the binary code of computers. By 1956, Newell and Simon had concluded that bits of information manipulated by a digital computer could stand for anything—not just numbers but words and features of the environment. The computer came to be seen as an all-purpose symbol manipulator that can interpret, recall, modify, and combine information and therefore carry out any operation which can be logically described.

Since then, computers have been programmed to carry out a wide variety of tasks, such as remembering, learning, reasoning, judging, and problem solving, which were previously considered uniquely human. Computers can play chess, forecast the weather, compose music, and diagnose and suggest treatment for physical disorders. Computers have been programmed to conduct sessions of psychotherapy that resemble those carried out by actual therapists.

Modern research on artificial intelligence has tended to take two different approaches. One school of research focuses on engineering problems and attempts to develop programs that can carry out tasks efficiently for the sake of getting them done, but a second group uses computers to simulate and test predictions about human information processing. One of the goals of artificial intelligence is to make computers more useful. The other is to understand more and more about the human mind.[5] The latter group of

[5] These two goals often conflict and limit research on artificial intelligence that is more directly related to psychology. Because military and industrial sponsors focus on efficiency, the "big money" goes into programs designed to make computer systems as efficient as possible, in order to overcome the limitations of human information processing. In designing an efficient calculator, for example, it would make little sense to saddle it with a working memory able to handle only 5 to 9 items at a time; and yet, without these restrictions it could never serve as a working model for human intelligence.

researchers attempt to break down the steps involved in information processing and describe them in minute detail. If sufficient details can be programmed into a computer that performs similar operations, then it provides a powerful argument about how human information processing occurs. Older linear models of the mind have been replaced by more dynamic models that use feedback and various self-correcting mechanisms.

If we view the nature–nurture debate in terms of the computer model, it is possible to draw a parallel between nature and nurture and the computer's hardware and software. The hardware is the computer itself—the physical unit, made up of wires and silicone chips, that processes information. Software refers to the programs and instructions that tell the computer what to do. If we now ask what is the relative contribution of each, the question is somewhat difficult to answer. Most computers, including home computers, have a potential far beyond their actual use. They are capable of feats undreamed of by their current owners. The world's most sophisticated computer, in contrast, is simply a very expensive paper holder without its instructions. The relative contribution of hardware and software can be seen as somewhat nonsensical in this context, because both are essential to the computer's operations.

Although the computer is a useful metaphor, important differences between computers and the human brain limit its application. First, there is the incredible complexity of the human brain. The human brain is the most complex organ known. The average brain has between 50 and 100 billion neurons. Some estimates range as high as a trillion. These neurons have an average of 2,000 to 3,000 connections. Schwartz (1988) has estimated that the analog arithmetic required to emulate the brain may be as high as 1,000,000 trillion operations per second. The largest supercomputers likely to be built in this decade, in contrast, will probably not attain speeds in excess of 1 trillion operations per second. These interconnections are not well understood, but even if they were, it would still be extremely difficult to design a computer of similar proportions. Nevertheless, building the hardware may be the *easiest* part of simulating the human brain (Waltz, 1988).

Second, the brain is not merely a system of neural connections. Evolution has provided not one but two means of transmitting information—one neural and one chemical—and there is a complex interplay between electrical and chemical events in the brain. A tiny cleft between neurons, called the synapse, converts electrical signals into a chemical code that releases substances known as *neurotransmitters.* These cross the synapse and link up with specific receptors on the other side. There are several dozen known transmitters and as many as 50 have been postulated. Many of these are directly linked to hormones associated with emotions and arousal. A second group includes the brain's natural opiates, endorphins. Many psychoactive drugs contain substances that bind to the same receptor sites and mimic the activity of neurotransmitters. Nicotine, for example, facilitates the transmission of acetylcholine; LSD blocks the transmission of serotonin; and chlorpromazine, which is commonly given to schizophrenics, interferes with dopamine and noradrenaline activity. The brain responds to subtle differences in the chemical structure of psychoactive drugs, even when they come from the same family.

A third major difference is based on the *plasticity* of the human brain. Unlike the computer's hardware, brain cells are alive and are altered by experience in a variety of ways. Pioneering studies by Rosenzweig and his colleagues beginning in the early 1960s found that rats raised in enriched environments, with various toys, wheels, and ladders, develop significantly more cerebral cortex and a greater number of glial cells that support and nourish the brain (Rosenzweig, 1984; Rosenzweig, Bennett, & Diamond, 1972). The increase in cortex size is not due to the formation of extra neurons but to greater dendritic branching and more synaptic contacts. Differences in experience affect mainly the cortex and there is little or no change within subcortical regions. Similar changes have been found in both young and older animals, although the latter require longer periods of exposure to achieve the same results (see Greenough, 1976). The results can be surprisingly specific. In the initial studies, the increase occurred primarily in those regions that serve visual perception. More recently, Greenough, Larson, and Withers (1985) trained rats to reach into a clear plastic tube with a specific paw to obtain pieces of chocolate chip cookies and later found that the opposite hemisphere (which controls the trained paw) had an increase in dendritic branches.

Experience also shapes the brain through a gradual pruning of those cells and synapses that are unnecessary or do not make proper connections. This process occurs in two ways. First, the brain of a newborn contains a large number of excess cells that eventually atrophy because they are redundant or injurious to normal functioning. Within months, more than half of these gradually degenerate and die. The brain goes through a period of "selective cell death" that may involve from 15 to 85 percent of the initial population, depending on the region involved (Cowan, 1979). Although the greatest loss of cells occurs during infancy, the process continues throughout childhood and into adolescence. In humans, cells continue to die within the frontal lobe until people are about 16 years of age (Huttenlocher, 1979; Kolb, 1989).

Second, most neurons initially generate more contacts than are actually needed. Many of these, and in some cases all but one, are later retracted. As Cowan (1979) has pointed out:

> Although many regions [of the brain] may be "hardwired," others (such as the cerebral cortex) are open to a variety of influences, both intrinsic and environmental. The ability of the brain to reorganize itself in response to external influences or to localized injury is currently one of the most active areas in neurobiological research." (p. 133)

Changes in brain structure based on experience are not limited to the number of connections. Recent research suggests that changes in the efficiency of synaptic connections occur as well (Lynch & Baudry, 1984).[6] Heightened sensitivity appears to occur through an influx of calcium, which triggers an enzyme (calpain) that "eats away" at the membrane of the postsynaptic junction, exposing more neurotransmitter receptors. More prolonged or repeated bursts of activity release larger quantities of calcium, producing long-lasting structural changes that may last for the duration of the cells. These changes also seem to occur primarily in the upper regions of the brain, such as the cortex, and there is little or no effect on lower brain regions, such as the brain stem and cerebellum.

All the evidence seems to support the view that the degree of flexibility varies with the region of the brain. As Gardner (1983) has pointed out, this high degree of *uncommittedness* reflects both the need for flexibility and the importance of the postnatal experience in determining the kind of connections that will ultimately take place. When it comes to highly complex activities, such as language, the young child can withstand considerable damage, such as the complete removal of a hemisphere, and still learn to speak relatively well. While variations in wiring would produce chaos in a computer, it is a necessary part of human evolution—allowing the brain to restructure itself on the basis of experience.

Structural differences between computers and the human brain translate into functional differences as well. The use of computers to simulate human intelligence has generally taken a piecemeal approach, whereby complex problems are broken down into discrete tasks that are then studied separately. But humans are always involved in many different activities. Even the simplest tasks, such as watching a sunset or opening a car door, require billions of operations at the same time. The kind of operations computers carry out are based on deductive reasoning, in which inferences are drawn from axioms and rules of derivation. Computers require explicit instructions to operate, but as Dreyfus and Dreyfus (1988) have pointed out, many aspects of human intelligence do not depend on explicit knowledge at all. They depend on "everyday know-how." Children, for example, learn to speak without explicitly knowing the rules of grammar. They later learn that the rules of language, such as pluralization, are not applied uniformly and learn the many exceptions to the rule. But they cannot explain *how* they form sentences or what makes one correct and another wrong. This tacit knowledge provides the background for much of what we do and think. From a phenomenological perspective, the whole purpose of psychology is to make this tacit knowledge explicit, but if this were ever accomplished, we

[6] This tendency was also noted by William James (1890) who said that, from a physiological point of view, *habit* is nothing but a new pathway formed in the brain. A pathway once traversed is traversed more easily a second time.

would literally know everything there is to know about the human mind. This does not mean we should stop trying. It simply suggests we may never create computers that think like humans: "Building an interactive net sufficiently similar to the one our brain has evolved may just be too hard" (Dreyfus & Dreyfus, 1988, p. 37).

Computers do a number of things extremely well, such as store and retrieve information, carry out complex mathematical and logical operations and so on. Humans have difficulty with these types of tasks because we are often unable to hold sufficient information in working memory to complete the operation. Whereas people have vast amounts of long-term memory, we are often poor at literal recall. We tend to "read through" specific symbols to derive the meaning or gist of the message. Research with bilingual speakers, for example, suggests that they frequently cannot even recall the language in which the information was conveyed.

There are other areas where computers are decidedly inferior to human beings. Ironically, computers are very good at complex tasks, such as mathematics and logical thinking, but extremely poor at "simple" tasks, such as vision and learning. The reason is that the explicit rules of logic and math can be easily programmed into a computer, but more "rudimentary" tasks are not well understood and are therefore difficult to duplicate. Unlike computers, people have an extremely limited capacity to process information. We are constrained by the number of items that can be held in working memory, the speed with which symbols can be manipulated, and our ability to retrieve information from long-term memory. To overcome these limitations, humans have developed various cognitive strategies such as rehearsing, organizing, and elaborating incoming information. Our limited capacity to process information has forced us to develop new ways of getting around these restraints. Individual items of information can be grouped together, or chunked, and processed as a unit. New items of information can be linked to older concepts. Humans have the ability to process the same information in different ways and cross-reference concepts and ideas. These abilities are based on overlapping semantic networks, which are also responsible for such things as humor, insight, and creativity.

Human information processing is also colored by emotions and motivational states. These factors are not simply tacked on. They are a vital part of human intelligence. They limit information processing in some cases, creating biases, distortions, and departures from rational models, but they also allow people to distinguish relevant and irrelevant information. Emotions provide a sense of purpose that motivates and directs ongoing activity. A significant aspect of human intelligence is the ability to selectively attend to only a small part of what is available. This allows people to focus on the most essential features while ignoring everything else.

Because many neurotransmitters are hormones, we can view the brain not just as the center for thought, but as a gland controlled by substances produced elsewhere in the body. Mood disorders, such as stress and depression, profoundly influence the way we think. Our desire and ability to process information is very much affected by interest, arousal, illness, and fatigue. The link between emotions and thought means that thought is not an activity confined to the brain. It involves the entire body.[7]

Skepticism about artificial intelligence is as old as cognitive psychology itself. As early as 1963, Neisser warned that there was a relative neglect of the context of cognitive development and that human thinking is strongly influenced by emotions and motivation. He later pointed out that a similar neglect was occurring among those involved in intelligence research (Neisser, 1976b). This does not mean that computers should not be used to simulate human information processing. Computers have proved their worth many, many times over. It simply suggests that, unless the processes are well understood, data derived from computer simulation should be viewed cautiously. The computer is a powerful metaphor and a useful model for some types of human information processing, but structural differences between the computer and the human brain suggest that they do not always function in precisely the same way. As Hurlbert and Poggio (1988) have pointed out, those who claim they can build computers that think like human beings have from now to eternity to prove themselves wrong. It is doubtful, however, whether the rest of cognitive science should wait. There is currently a great deal of research that merely needs to be reviewed and organized.

Neurological evidence has been used in the past to support a rather narrow view of biological determinism. In fact, at one time, it was widely believed that there was a direct relationship between intelligence and the overall size of the brain. One of the leaders in this area was the French neurosurgeon, Paul Broca (1824–1880), who was also responsible for the discovery that the ability to speak is localized within the left frontal lobe. Broca's view was largely abandoned when it was discovered that Germans had larger brains than the French and that the brains of executed criminals were often bigger than those of eminent men of science. Broca's own brain weighed only 1,424 grams, a bit above average but not much (Gould, 1981).

A slightly more sophisticated version views specific abilities as being due to differences in particular regions of the brain. The genetic position suggests that genetic differences produce variations in brain structure that lead to individual differences in mental ability. This can be used to explain why some traits, such as musical ability, seem to run in families. The recent

[7] There is no reason to confine mental activity to the limits of the skin. When two people talk and share their ideas, they are engaged in a mutual thought process. Thought begins as a social activity. Only later do children learn to converse with themselves and still later to think silently without speaking out loud (see Chapter 7). The meaning of a message depends on the social context. Words are qualified or even disqualified by nonverbal cues occurring at the same time. Some aspects of conversation, such as sarcasm and teasing, are completely incomprehensible unless the context is taken into account (see Collier, 1985).

neurological evidence, however, suggests that the latter part of this causal relationship sometimes points the other way. People develop skills and abilities by adapting to their environment and these adaptations alter pathways and structures within the brain. When it comes to human intelligence, the critical question is not the relative contribution of hardware and software but the role of genetics and learning in each.

THE "NEW LOOK" IN INTELLIGENCE RESEARCH

Given the overlapping interests, it was merely a matter of time before the areas of psychometrics and cognitive science converged. This has created what Keating (1984) has called the "new look" in intelligence research and a level of activity that would have been difficult to predict two decades ago. Unlike psychometrics, which was concerned primarily with the structure of intelligence, the "new look" examines the processes and subprocesses underlying actual performance. The emergence of cognitive psychology as the dominant psychological approach is already beginning to force a reconceptualization of what it means to be intelligent and an expansion of cognitive psychology into areas previously dominated by mental testing.

The range and diversity of this research defies a quick summation, but some examples may be useful. Early research by Earl Hunt (1976, 1978) focused on processes involved in *verbal comprehension.* The general theory underlying this research is that information processing occurs in a series of stages beginning with the sensory input and ending with permanent storage in long-term memory. Information from the environment is compared with previous material in long-term memory. If it is recognized, a name or label is applied and it goes into short-term memory where it is retained for a few seconds. Meaningful information from short-term memory then passes into an intermediate-term memory that may last for minutes or even hours. At each stage, the ability to process information is determined jointly by the knowledge already in long-term memory and a number of lower-level mechanistic processes that determine the speed and accuracy of information processing.

Robert Sternberg (1977) has carried out a similar line of research on the steps involved in *analogical reasoning.* His theory involves six components: encoding, inference, mapping, application, justification, and response. The general model can perhaps best be illustrated by using an example of a verbal analogy used by Sternberg (1985a):

> Lawyer is to client as doctor is to ____________.

During the *encoding* phase, important aspects of each item are noted and compared with information in long-term memory. The term *lawyer* might involve the recognition that a lawyer is a person who provides legal services. During the *inference* stage, the initial pair of items are compared for similarities and differences and some kind of relationship(s) is established. In this example, the relationship is based on the fact that a lawyer provides legal services and a client receives them. *Mapping* is similar, but this time the focus is on the first and third item (lawyer and doctor) and might include the recognition that both lawyers and doctors provide professional services. *Application* involves applying a specific rule inferred during the previous stages to the third term to create a hypothetical match: A patient receives medical services from a doctor in much the same way that a client receives legal services from a lawyer. *Justification* is a checking operation that leads to the rejection of inappropriate alternatives and selection of the correct *response.* Geometric and figural analogies are solved in a similar way, but the focus is on physical attributes, such as size or shape.

Sternberg has used a variety of procedures to test the internal and external validity of his theory. Internal tests frequently involve the use of mathematical models to create a reaction-time equation based on the number of processes involved. These are then compared across subjects and across tasks to see how well they predict actual performance. Sternberg also uses precuing procedures to eliminate some of the steps. Subjects may be allowed to study the first two items as long as they wish and then solve the problem. The reaction time necessary for solution in this condition does not include the time necessary for encoding the first two items and inferring their relationship. External validity was assessed by correlating the performance of individual subjects with psychometric measures of reasoning ability. Sternberg found a sizable positive correlation between encoding time and reasoning ability scores—that is, subjects who perform well on general tests of reasoning ability spent *more* time encoding individual items.

A third line of research focuses on *problem solving* (e.g., Chi & Glaser, 1985; Resnick & Glaser, 1976). Problems occur in situations in which a person is attempting to reach a goal and there is no immediate solution. Problem solving involves three general processes—problem detection, feature scanning, and goal analysis. The person begins by assessing the problem and searching long-term memory for a possible solution. If a solution can be found, it is simply applied and tested for success. If not, the task must be searched for relevant cues, which are then compared with items in long-term memory. More typically, a partial solution is found, and this information must be kept temporarily in working memory as the process continues. If the initial goal cannot be reached, then the problem must be redefined and a new goal established.

Research on problem solving typically uses puzzle problems because they require little background knowledge but some studies have also been carried out by comparing novices and experts in particular domains. De Groot (1965), for example, compared chess masters and less experienced players and found no evidence that chess masters considered more moves or searched further ahead. The only difference was that chess masters recognized and considered better moves. Chase and Simon (1973) later found that part of their superior performance was due to the ability to recognize meaningful patterns and treat them as a unit. As in other areas, the ability to solve problems depends on both the nature of the task and the amount of previous knowledge.

These brief accounts of research on verbal ability, analogical reasoning, and problem solving help illustrate some of the general features of the information-processing approach to intelligence. Complex tasks are broken down into basic components that are then studied separately. The extent of the breakdown varies enormously. Hunt's research focuses on extremely elementary skills, such as the speed of lexical access or the ability to manipulate verbal information, whereas research on problem solving examines more general skills, such as feature scanning and goal analysis. Research techniques range from comparative studies of people with high and low ability in a particular area to sophisticated mathematical modeling and computer simulation. Individual differences occur because people vary in the speed and accuracy with which they apply individual operations, the amount of background knowledge, and use of higher order skills that select and coordinate ongoing activities (see Pellegrino & Glaser, 1979).

Sternberg (1985a) has recently developed a *triarchic theory of intelligence* in an attempt to integrate the many diverse versions of mental ability. His theory consists of three subtheories addressing different aspects of intelligence. The *experiential subtheory* defines intelligence as the ability to adapt to new situations and the automatization of processes and subprocesses until they become habitual. The automatization of basic processes allows more complex tasks to be carried out because the operations run smoothly and can be taken for granted. According to Sternberg, the best tests of intelligence occur, not in run-of-the-mill situations that are frequently encountered, but in new situations that challenge a person's ability to cope. The experiential view shows one of the reasons it is so difficult to compare people from different subgroups and cultures. To know that a task is truly novel, the researcher must know what tasks the person has previously confronted. Even when tests measure the same type of skills, they will very likely differ in terms of the degree of automatization and the level of novelty.

The *componential subtheory* specifies the structures and mechanisms underlying intelligent behavior. It consists of three major components—metacomponents, performance components, and knowledge-acquisition components. Metacomponents are higher order executive processes for planning, monitoring, and evaluating ongoing events. The performance components execute plans formulated by metacomponents and include subprocesses such as encoding, combining and comparing information, and responding. The knowledge-acquisition components are used to obtain and assimilate new information. They selectively encode and combine new material and compare it with information previously acquired.

The *contextual subtheory* attempts to show how intelligence is modified by the larger cultural context. It has three aspects—(1) adaptation to the present environment, (2) selection of more favorable conditions, and (3) shaping the present environment so as to achieve a better fit with the person's skills, values, and interests. This part of the theory is "relativistic" in that it recognizes that demands placed on people vary from individual to individual and from culture to culture.

Sternberg's theory has several advantages and disadvantages. It is broad enough to encompass a great deal of empirical research but is sufficiently abstract to undergo considerable modification. It should be possible to add additional components and deemphasize others without seriously damaging the overall structure. There is no claim that the structure of the theory mirrors the structure of the mind. It merely identifies three central aspects of intelligence.

The most serious weakness (from the present point of view) is that the "contextual" subtheory is hardly a contextual theory at all. It merely identifies three broad coping strategies—adaptation, selection, and shaping—that modify intelligence, but it does not describe *how* intelligence operates within different contexts or how context helps shape the nature and varieties of intelligence. There is also a tendency to treat adaptation, selection, and shaping as mental processes occurring within an individual. This is a restriction that Sternberg (1985a) regards as "necessary in order to keep the conception of intelligence within mental (and manageable) bounds" (p. 329).

A third weakness is a tendency to see intelligence as pretty much the same in all cultures. This is true at a very general level. All people encode, process, and retrieve information; and all people plan, monitor, and correct ongoing behavior. It is unclear, however, whether the terms *encode, process,* and *retrieve* refer to the same processes when applied to different cultures or whether people from different cultures plan, monitor, and correct ongoing activities in the same way (see Galotti, 1989). It is important to find a level of analysis where differences count. Sternberg (1985a) is well aware of these limitations. He points out that "the contextual subtheory is the least elaborated of the three subtheories, and it is also the one most in need of further research" (p. 328).

The "new look" in intelligence research represents a considerable advance over the old psychometric approach, but there are still several limitations and weaknesses. The first is based on the *selection* of cognitive tasks. Sternberg (1985a) has pointed that all previous theories of intelligence, both cognitive and psychometric, "suffer from a common flaw: They are post hoc. Investigators have started with a class of tasks and then claim that intelligence is whatever it takes to do well on those tasks" (p. 67). He goes on to suggest that task selection in cognitive psychology has been even less theoretically motivated and is at least as haphazard as that inspired by psychometric research. There is still considerable disagreement over what is meant by the concept of intelligence when people are asked to list features associated with intelligence or to identify traits possessed by intelligent people. Professionals working in the area are somewhat more likely to include motivational factors, such as "studies hard," "is persistent," and "gets involved," whereas laypeople place a somewhat greater emphasis on social skills (Sternberg, Conway, Ketron, & Berstein, 1981).

The second problem is based on *validity.* Since there is little agreement over the concept of intelligence, there is considerable disagreement about what criteria can be used to test current models. Cognitive models draw empirical support from performance on "canned" tasks that may lack external or ecological validity (Sternberg, 1985a). There is also a vicious circle in intelligence research whereby investigators who reject older models of general intelligence fall back on IQ tests for lack of a better criterion (U. Neisser, 1978, 1983).

A third problem is one of *omission.* Gardner (1985) has pointed out that a paramount feature of cognitive science is "the deliberate decision to de-emphasize certain factors which may be important for cognitive functioning but whose inclusion at this point would unnecessarily complicate the cognitive-scientific enterprise. These factors include the influence of affective factors or emotions, the contribution of historical or cultural factors, and the role of the background context in which particular actions or thoughts occur" (p. 6). The focus on "central" processes, which was to some extent an overreaction to the extreme "environmentalism" of behaviorism, has tended to place thought and intelligence "within the head" of the individual and has downplayed the role of external conditions and culture. As Sternberg (1985b) has pointed out:

> Theories in the differential and cognitive tradition have almost never been sensitive to the interface between intelligence and the cognitive context in which it is exercised; yet many psychologists, anthropologists, and others have doubted whether we can fully understand intelligence except in relation to the contexts in which it develops and is exercised (see, e.g., Berry, 1974, 1981, 1982; Charlesworth, 1976, 1979; Dewey, 1957; Keating, 1984; Laboratory of Comparative Human Cognition, 1982, 1983; Neisser, 1976 [a], 1978; Sternberg, 1984; Sternberg & Salter, 1982). I do not believe that the failure to consider contextual variables necessarily renders the differential and cognitive theories wrong, but it does render them incomplete. (p. 23)[8]

Sternberg (1984) recognizes that intelligence varies with context, but he does not go far enough in describing the specific effects.

[8] All the references cited in this quotation are listed in the bibliography at the back of this book.

If intelligence can be tentatively defined (à la Sternberg) as the ability to adapt to new situations and learn from experience, then a much greater emphasis must be placed on motivation and emotions. Professionals commonly recognize that these factors are an important aspect of intelligence. It is also clear that individual and cultural differences in intelligence cannot be properly understood without a clearer understanding of the role of the environment. Intelligence can be seen, at least in part, as a process whereby an individual acquires skills during the course of adaptation that are then practiced until they become automatic. The particular combination of skills, however, depends on the context, and therefore, intelligence research must reach out and include the specific contributions of a number of additional disciplines that may be broadly defined as "social psychology."

PART TWO

Motivation

CHAPTER 3

Achievement Motivation

This chapter examines the research on achievement motivation, beginning with Max Weber's concept of the Protestant work ethic and Henry Murray's more general theory of psychological needs. Childhood antecedents and cognitive consequences are explored, as is the related concept of "fear of success," which was originally formulated to explain relatively lower levels of achievement motivation among women. The chapter closes with a review of the more recent attempts to explain achievement motivation in terms of attribution theories. Gender differences are again discussed, along with Carol Dweck's division of children into those who are "mastery oriented" and "helpless."

Achievement motivation is a useful place to begin when examining social origins of mental ability because there is a long history of research and a considerable body of data based on both experimental and cross-cultural studies. There is also compelling evidence that achievement motivation is an acquired trait in which genetic factors play a relatively minor role. Variations in achievement motivation are widespread across cultures and changes within cultures occur much too rapidly to be attributed to heredity. Historically, high levels of achievement motivation tend to precede periods of economic prosperity, whereas reduced levels are associated with economic decline and many cultures have shown marked shifts in achievement motivation within a single generation (McClelland, 1961).

Achievement motivation can be defined as a conscious or unconscious desire to perform well and reach high standards of excellence (McClelland, 1961). It is positively related to academic performance even when people score equally on standard intelligence tests (Morgan, 1952; Ricciuti & Sadacca, 1955). Achievement motivation is acquired early and appears to remain relatively stable throughout life. Variations in achievement motivation have been found in children as young as five (McClelland, 1958), and preschool children who strive to master intellectual and cognitive tasks tend to maintain high levels of achievement motivation throughout high school and into early adulthood (Moss & Kagan, 1961). Achievement motivation affects not just our level of aspiration but how we think and process information. It is positively related to a person's willingness to take reasonable risks, delay gratification, and sustain activities in the face of failure. On complex tasks, those high in achievement motivation start at the same level as those with low achievement motivation but do progressively better as they proceed (McClelland, 1961). Many of the original studies were conducted during the "Golden Age" of research on achievement motivation, which occurred during the 1950s and 1960s. During the 1970s, there was an increasing tendency to explain achievement motivation in cognitive terms either through expectancies based on locus of control or through attribution theories. The research on attribution models of achievement motivation will be discussed in the final section of this chapter, whereas locus of control will be the focus of Chapter 4.

THE GENERAL THEORY

The theory of achievement motivation can be traced back to Max Weber's description of the Protestant work ethic. Max Weber (1864–1920) was a German social theorist writing at the turn of the century, largely in response to what he perceived as a strict view of economic determinism.[1] According to this view, attitudes and beliefs are shaped by economic conditions, and changes in economic institutions produce noticeable shifts in philosophy, religion, law, and even consciousness itself. Weber argued that economic change is embedded within a larger cultural context and that changing attitudes brought on by religion or ideology often occur prior to economic reform. Although he provided a comparative study of world religions to support his case, his classic work, *The Protestant Ethic and the Spirit of Capitalism* (1904–1905/1958), focuses on changes occurring during the Protestant Reformation.

Feudal society before the Protestant Reformation was based primarily on self-sufficient peasant farming, supplemented by domestic industries and crafts in small towns. The traditional worker in feudal society worked only as much as necessary to produce goods for himself and his family, after giving a portion to the church and his feudal lord. If new, more efficient modes of production were introduced, workers responded by working less rather than producing more. Each estate was a self-sufficient unit producing primarily for itself. The accumulation of wealth, while widely practiced, was not sanctioned by the Catholic Church and was, at least in principle, discouraged.

The development of Protestantism changed these values and produced a new type of individual uniquely suited to a capitalist economy. The Protestant

[1] Weber was reacting to the view commonly attributed to Karl Marx (1818–1883) and his followers. Marx did not actually hold a strict view of economic determinism and was quite critical of it in his early work. He felt that people made their own history but not under circumstances of their own choosing. Weber's opposition to Marx developed partly because many of Marx's early writings, in which he developed his thoughts on the relationship between consciousness and society, were not published until after Weber's death (see Giddens, 1971).

leader, John Calvin (1509–1564), advanced the notion of a *calling,* in which each person received a profession from God and had a moral obligation to develop it as much as possible. He also stressed the doctrine of *predestination,* whereby people are chosen in advance for either salvation or damnation. The belief in predestination is simply a logical extension of a belief in an almighty God. A truly all-powerful God is not affected by the prayers and behavior of individuals. The doctrine of predestination changed people's attitudes toward the church and each other. If a person could not judge others by their deeds, it was impossible to have confidence in an established religion because even the local priest, who was previously responsible for salvation, might not be among the elect. Calvin preached the doctrine of a "hidden" church consisting of those who were elect, but whose identity was known only by God. Followers could not even become too attached to family members because they too might be condemned to hell. Each person stood absolutely alone, without intermediaries, before God.

The concept of a calling and predestination changed people's attitude toward work as well. It might be assumed that a belief in predestination would lead to fatalism and resignation (and this has been the response in many religions throughout the world), but the Protestants took success as a sign of salvation. Success was seen as an indication that a person was truly saved. Success could not bring salvation since that was determined in advance but it could be used to reduce the gnawing uncertainty of being condemned. The belief in predestination produced a devastating sense of isolation and uncertainty, the only escape from which was hard work. Protestantism raised work to the level of a moral obligation and made idleness a sin.

Protestantism not only encouraged the accumulation of wealth as a means for reducing the uncertainty of salvation but also discouraged people from spending money on themselves. Puritanism, in its original form, opposed any form of spontaneous enjoyment or self-indulgence—games, art, theater, or any ostentatious display of wealth. The result was a hard-working individual, continually striving to earn money, but unable to use it for personal enjoyment. The only alternative was to reinvest the earnings and make more money. In this way, money became disassociated from its previous function of buying goods and simply became a means for measuring success. Such an individual, according to Weber (1904–1905/1958), "gets nothing out of wealth for himself, except the irrational sense of having done his job well" (p. 71).

The pursuit of wealth is not unique to capitalism. People in all ages have valued wealth and have taken it whenever they found an opportunity. What distinguished capitalism, at least in its initial stage, was the rationally planned acquisition of wealth based on hard work, the pursuit of profit, and reinvestment. This produced not only a profound economic change, leading ultimately to the industrial revolution, but a profound psychological change as well. The result was an obsessive individual continually striving to produce more and more but unable to enjoy the fruit of his or her labor because of its association with sin and self-indulgence. The need to work hard and succeed is certainly not a fundamental part of human nature. It is relatively rare among cultures even today. American entrepreneurs who go abroad often complain of the inability to recruit a dependable, well-motivated work force that is willing to fit into a scheduled routine. What they fail to recognize is that this capacity is itself acquired and is the product of a unique set of cultural conditions.

Research on achievement motivation was also strongly influenced by Henry Murray's (1938) theory of personality. At a time when most psychologists were developing theories with greater and greater simplicity, Murray provided an extensive taxonomy of human motives that he considered necessary to account for the complexity of human behavior. Needs are relatively stable latent tendencies to behave in a particular way under the appropriate conditions. For Murray, human behavior is goal-directed, and the most important thing we can know about a person is the direction and intensity of his or her aspirations.

Motives can be divided into two groups—*primary needs,* which are linked to specific changes within the body, and *secondary needs,* which have no clearly defined physiological source and are acquired, or at least modified, during socialization. Murray was primarily concerned with secondary needs and drew up a list of 28 basic drives, such as the need for achievement, autonomy, play, and affiliation, which guide and direct most human activity. Needs may be conscious or unconscious. Conscious needs are socially acceptable and may be expressed more or less directly, whereas unconscious ones are unacceptable and generally manifest themselves in dreams and fantasies. Not only are people unaware of some of the motives that influence their behavior, but they actively ward off or repress knowledge of others.

Because some needs are unconscious and socially unacceptable, they cannot be assessed through direct questions or standard questionnaires. Murray developed the *Thematic Apperception Test* or *TAT* as an indirect measure. The TAT consists of 29 ambiguous pictures and a blank card. Subjects are shown each card one at a time and are asked to describe what is happening, what led up to the event, what the characters are thinking and feeling, and how the event will turn out. Because the stimuli are ambiguous, subjects are free to respond any way they choose, and their descriptions are taken as an indication of their needs. One of Murray's needs, the need for achievement, became the focus of research by David McClelland and John Atkinson at Wesleyan University during the postwar period (McClelland & Atkinson, 1948; McClelland, Clark, Roby, & Atkinson, 1949) and a summary of their early work was presented in the book, *The achievement motive* (McClelland, Atkinson, Clark, & Lowell, 1953). Four cards from the original TAT were retained, and procedures similar to those of Murray were used to elicit responses.

During the late 1950s and early 1960s, research on achievement motivation went in two different directions. McClelland (1961) and his associates at Harvard focused primarily on the social origins of achievement motivation and its role in economic development. They carried out extensive cross-cultural research and compared levels of achievement motivation in previous cultures during periods of economic prosperity and decline. Atkinson and his associates at the University of Michigan became increasingly concerned with developing and testing predictions based on various mathematical models (Atkinson, 1957, 1983; Atkinson & Feather, 1966; Atkinson & Raynor, 1974). Atkinson's (1957) early model differentiated between two different aspects of achievement motivation—the motive to approach success and the motive to avoid failure—and attempted to account for the joint effects of achievement motivation conceived as a personality variable and the degree of challenge offered by the situation.

CHILDHOOD ANTECEDENTS

One of the changes brought on by the Protestant Reformation was a changing attitude toward literacy. Since Protestant leaders felt that people could not rely on the established clergy, they encouraged their followers to read the Bible and seek divine guidance directly. McClelland (1961) has pointed out that this literacy training stressed early mastery in the home and believes that it is reasonable to interpret Weber's treatment of the relationship between Protestantism and capitalism as a revolt within the home.

A study by Winterbottom (1958) conducted in 1953 suggested that individuals who are high in achievement motivation undergo early independence training and are expected to develop skills at a relatively early age. She interviewed mothers with 8-year-old sons who were high and low in achievement motivation.[2] Mothers with sons who were high in achievement motivation expect their sons to master such activities as knowing their way around their neighborhood, trying hard without asking for help, making friends among children their own age, and doing well in competition. Mothers with sons low in achievement motivation placed more restrictions on their sons and did not want them to play with other children without their approval or make important decisions by themselves. McClelland, Rindlisbacher, and de Charms (1955) used items identified by Winterbottom to study Protestant and Irish and Italian Catholic parents living in New England. They found that each group encouraged self-reliance, but Protestant mothers expected mastery at a significantly earlier age.

[2] The bulk of the early research focused on male children and college students because it was discovered quite early that females do not show the same trends. Sex differences in achievement motivation will be discussed in more detail in the next section of this chapter.

Similar differences have been found by Rosen and D'Andrade (1959) when parents and sons were observed within the home. Parents of boys high in achievement motivation typically set high standards for performance and rewarded their sons for doing well. They showed greater warmth in general but gave special approval for success. Fathers of high-achieving boys encouraged independence by giving general rather than specific instructions. Crandall (1969) has found that mothers of children high in achievement motivation tend to reward their children's efforts to achieve frequently and spontaneously, even when the children are not seeking approval.

Moss & Kagan (1961) provided evidence that there is an optimal age for producing achievement motivation. Mothers with extremely high achievement motivation may make excessive demands at an early age and actually lower achievement motivation in their sons (McClelland, 1961). Rigid and authoritarian fathers also tend to have sons who are more dependent and low in achievement motivation. McClelland (1961) has gone so far as to suggest that achievement motivation may *increase* in authoritarian societies during periods of war and revolution because fathers are taken away from the home and often do not return.

Teevan and McGhee (1972) studied mothers of children who were high and low in *fear of failure* and found that mothers with children high in fear of failure also expected self-reliance and displays of success at a relatively early age. They were more inclined, however, to ignore successful performances and punish unsuccessful ones. Mothers of children with low fear of failure tended to reward satisfactory performances but were less inclined to punish unsatisfactory ones.

The two motives—achievement motivation and fear of failure—thus seem to have slightly different childhood antecedents. Parents with children high in achievement motivation tend to set high but realistic standards and reward success, whereas those with children high in the fear of failure tend to expect levels of performance beyond the child's ability and are more inclined to punish or criticize unsuccessful performances. Those high in achievement tend to come from democratic homes with nonauthoritarian fathers and warm, affectionate mothers, whereas those high in the motive to avoid failure are raised in homes where parents stress success very early in the child's life and punish failure with criticism and disapproval. Those with a high motive to succeed tend to seek success and anticipate positive consequences such as self-satisfaction and pride, whereas those who are motivated to avoid failure tend to avoid tasks in which failure is possible because it is associated with negative consequences, such as shame, rejection, and humiliation.

The independence of these two motives is also shown by the fact that there is little or no correlation between them (e.g., Atkinson & Litwin, 1960; Brody, 1963; Mahone, 1960). It has been relatively easy to find individuals who score high or low on both motives—that is, high in both achievement motivation *and* the fear of failure or low in both achievement motivation *and* the fear of failure. Some studies (e.g., Karabenick & Srull, 1978) have actually found more subjects high or low on both motives than those high on one and low on the other when a mean split was used to divide subjects into groups. One could infer that those high on both motives had parents who set and reinforced high standards but were hypercritical at the same time. Those low in both motives were either neglected or had parents who deemphasized success. Neglected children tend to become independent at a relatively early age but they do not always develop high standards (Rosen, 1959).

Numerous cultural and subcultural differences in achievement motivation have also been found. The relationship between early independence training and achievement motivation has been confirmed in studies of nonindustrial societies (Child, Storm, & Veroff, 1958; McClelland & Friedman, 1952). Early, rigorous independence training is associated with high achievement motivation in adults. Where independence training is late or casual, achievement motivation is typically low. There are exceptions, however. DeVos (1965) has found a different pattern in Japan. Although Japanese parents stress achievement and success, they also place a strong value on social and family ties and downplay independence and self-reliance. Chinese and Greek cultures also treat success as a family enterprise, and some have suggested that this has prevented them from developing large businesses based on impersonal labor. It may be that the close association between achievement motivation and independence is partly a remnant of the Protestant ethic described by Weber.

Many contemporary investigators have focused on *religious* differences in achievement motivation. The previous study by McClelland, Rindlisbacher, and de Charms (1955), which contrasted Protestant and Catholic children, has already been mentioned. Lenski (1958) carried out an extensive survey of Protestants and Catholics in the Detroit area and found Protestants were more inclined to stress independence, individual initiative, and mastery of the environment. Fewer Catholics believed working-class children had a good chance of becoming prosperous or that ability was more important than family connections. Rosen (1959) studied more than 400 mother–son pairs representing six ethnic groups in New England and found higher levels of achievement motivation among Protestants, Greeks, and Jews than among Italians, blacks, and French-Canadians. There were also significant differences based on socioeconomic class, however. Middle-class boys scored higher than *either* upper-class or lower-class groups.

Numerous other studies have shown social-class differences (e.g., Douvan, 1958), and the consistency of such data has led McClelland (1961) to conclude that social class is even more important than religion. He believes that social-class differences can be traced in part to the parents' occupation. Blue-collar workers often engage in simplified tasks in which they contribute only a small part. They have little opportunity to develop skills or individual initiative, and their chances of being promoted or dismissed are often beyond their control. Middle-class occupations generally require more education, long-range planning, and hard work without immediate gratification, and these values are passed on to the children.

One advantage of the content-analytic procedure used to assess achievement motivation in projective tests is that it can be applied to *any* written material—past or present. McClelland (1961) has studied the level of achievement motivation among writers in ancient Greece, industrial England, and Spain during the Middle Ages and has found a general tendency for high levels of achievement motivation to precede periods of economic expansion and drops in achievement motivation just prior to periods of economic decline. The relationship between prosperity and achievement motivation may help to account for the cyclical nature of economic development and ultimately the rise and fall of great nations and empires.

Although the level of achievement motivation may fluctuate within cultures, there is considerable evidence that it remains relatively stable within individuals. As noted previously, differences in achievement motivation have been found in children as young as five (McClelland, 1958), and variations among preschool children tend to correlate with those in high school and early adulthood (Moss & Kagan, 1961). Longitudinal studies have also shown that teenagers had roughly the same level of achievement motivation when tested six years later (Kagan & Moss, 1959).

There is evidence, however, that achievement motivation can be increased in some cases. McClelland (1978) trained businessmen in a village in India to think and act like people with high achievement motivation. When this group was compared with a comparable group in a neighboring village, those who were trained started more new businesses and employed over twice as many new people during the next two years. Kolb (1965) had underachieving boys with high IQs (above 120) participate in an achievement change program at summer camp. Those participating were subdivided by socioeconomic status and compared with a control group. Kolb found that school grades improved for all groups during the following year, but the improvements were maintained only among those in the experimental group from upper-class homes. Presumably, returning lower-class children to lower-class neighborhoods in which achievement values are not widely shared nullified the gains made during the training session. This suggests that achievement motivation is not just produced by social conditions but is *maintained* by social conditions occurring at the same time.

High achievement motivation does not guarantee success, however. Achievement motivation is often high among people in underdeveloped countries who lack the natural and technical resources necessary for economic growth. Prosperous countries, on the other hand, may maintain high living standards (for a while) in spite of significant drops in achievement motivation. Some groups value hard work, self-reliance, and self-sacrifice, but their conservative beliefs cause them to avoid readily available technical innovations, making them *less* productive than their more innovative neighbors. There is a complex interplay between achievement motivation and actual performance that varies from group to group.

But, overall, high achievement motivation is an asset. Those high in achievement motivation differ from those low in achievement motivation in a number of significant ways.[3] During the 1950s, 1960s, and early 1970s, Atkinson and his associates (Atkinson, 1957; Atkinson & Feather, 1966; Atkinson & Raynor, 1974) developed a number of mathematical models to predict behavior, based on the joint effects of the motive to succeed and the fear of failure. They found that those who are relatively high in achievement motivation tend to prefer tasks that are moderately difficult, whereas those with low levels of achievement motivation select tasks that are either overly difficult or extremely easy. This occurs both under controlled laboratory conditions and in real life. Students high in achievement motivation are more likely to choose occupations that correspond to their level of ability, whereas those high in the fear of failure aspire to occupations that are either overly difficult or too easy (Mahone, 1960; Morris, 1966). A similar pattern has been found for choice of majors among male college students (Isaacson, 1964). Students relatively high in achievement motivation chose courses that were moderately difficult, whereas those low in achievement motivation chose courses that were either very difficult or too easy.

The selection of overly difficult tasks seems to be primarily a defensive strategy designed to protect self-esteem and avoid social disapproval in the face of failure. Attempting a difficult task brings social approval without the corresponding threat of failure, since almost everyone is expected to fail when the task is extremely difficult. The selection of an extremely easy task virtually ensures success, whereas the selection of a difficult task virtually guarantees failure. Therefore, people with low achievement motivation never have the opportunity to develop their full potential or evaluate their actual level of ability.

Another prediction that has been made and verified in a variety of studies is that people high in achievement motivation are more persistent in the face of failure. Not only do they come to situations with a stronger belief in their own ability (Pottharst, 1955; Feather, 1965), but moderately easy

[3] To avoid unnecessary repetition, those high in the motive to succeed and low in the motive to avoid failure will simply be referred to as "(relatively) high in achievement motivation," whereas those low in the motive to succeed and high in the motive to avoid failure will be described as "(relatively) low in achievement motivation" or "high in the fear of failure," depending on what aspect is being stressed. When a mean split is used to assign college students to groups, it is by no means clear that the motive to avoid failure actually exceeds the motive to succeed. These subjects are simply relatively low in the motive to succeed (i.e., below average) and relatively high in the motive to avoid failure (i.e., above average).

tasks actually become more attractive once failure occurs, because they now appear to be more difficult. People high in achievement motivation come to see these tasks as a challenge and increase their effort to master them. If there is a choice of tasks, on the other hand, then those relatively high in achievement motivation are less likely to pursue tasks that are either extremely difficult or extremely easy (Feather, 1961, 1963; Moulton, 1965). This again ensures a closer match between the task and the person's level of ability, leading to an optimal development of whatever skills they do possess.

There is also evidence that those high in achievement motivation prefer goals that require a large number of steps to complete, whereas those low in achievement motivation tend to avoid tasks with a large number of steps in the series. Raynor (1974) devised a new formula that he used to compute the relative level of achievement motivation when more than one step is necessary to complete a goal, such as obtaining a medical degree or conducting research on a particular project. People who are motivated to succeed become increasingly more motivated as the number of steps increases, whereas those high in the fear of failure become increasingly more inhibited. The new formula also predicts that those with high levels of achievement motivation will choose tasks, work hard, and perform better on tasks that are relatively easy when there are many steps in the sequence. The choice of an easy task makes later steps more likely and increases the probability of actually reaching the final goal.

Raynor (1974) has used these differences to paint a general picture of those high in achievement motivation as they pursue their careers. Such people initially select a relatively easy task where success is highly probable. As confidence and ability increase, new tasks are selected that still appear easy to the person as an "expert" in the area but seem difficult and demanding to those on the outside. Successful individuals continually strive toward moderately easy and realistic goals but their level of performance is so superior to the norm that they appear to be "overachievers" to those looking on. It is important to note that highly motivated people are not merely testing their ability, although this is certainly an important aspect of their performance. They are developing skills by mastering challenging activities that will allow them to perform better on similar tasks in the future.

FEAR OF SUCCESS

The McClelland-Atkinson model of achievement motivation focuses primarily on achievement in males. In male subjects, task selection, persistence, and various other aspects of achievement-related behavior can be predicted reasonably well if the researcher knows the relative strength of the motive to succeed, the motive to avoid failure, and the perceived probability of success (Atkinson & Feather, 1966). Attempts to replicate these findings with women, however, have not been very successful. It is now known that actual achievement is based on a number of additional factors, such as the desire to please or avoid criticism, but these were not incorporated into the original theory in any systematic way. In a massive work on achievement motivation (Atkinson, 1958), only a footnote was devoted to achievement motivation in women, but within that footnote sex differences were described as "perhaps the most persistent unresolved problem in research on Achievement" (cited in Horner, 1968).

In the late 1960s, Matina Horner (1968) submitted a doctoral dissertation at the University of Michigan in which she introduced a third motive—the *fear of success*—to help account for these inconsistencies. She derived her theory from the observation that highly competent women often experience conflict when confronted with standard achievement tasks and reduce their effort just when they are on the threshold of success. Such women want to succeed but they also fear the social disapproval and loss of femininity associated with success, particularly on traditionally male tasks. They are threatened by success because outstanding performance is unconsciously associated with low femininity and consequently social rejection. The motive to avoid success was conceived as a relatively stable, latent disposition acquired early in life. It is not the same as the wish to fail. The motive to avoid success is based on perceived negative consequences associated with success, whereas the wish to fail implies positive consequences based on failure. Horner suggested that high fear of success is characteristic of highly able, achievement-oriented women who have the potential to succeed but fear it at the same time.

To explore these ideas, Horner (1968) asked women to respond to the following hypothetical situation: "After the first-term finals Anne finds herself at the top of her medical school class." Male subjects were given the same description but with the name "John" and the proper pronouns substituted. Fear of success was scored as present if the description included negative consequences associated with success, activities away from future success, direct expressions of conflict about success, denial of effort or responsibility, or bizarre and inappropriate responses. Horner found that fear of success imagery was dramatically higher among women—62% of her female subjects wrote stories with some fear of success imagery, and such stories were more common among women in honors programs. A common theme was social rejection. In one story, Anne deliberately lowered her academic performance, helped her boyfriend Carl achieve better grades, dropped out, married Carl, and raised a family, while Carl went on to medical school. Only 9% of male subjects expressed fear of success. The remainder showed strong positive feelings toward success and a belief that success would bring other rewards.

Horner's study produced immediate excitement among those working in the area. She was one of the first to call attention to the ambivalence associated with success and the perceived "cost" of success for women. Because her theory offered an additional environmental explanation for sex differences in achievement and a way out of the impasse provided by the previous literature, it generated a number of follow-up studies. Zuckerman and Wheeler (1975) have provided a useful review. They reviewed 15 studies carried out during the 1960s and 1970s and found that the differences between men and women were less robust than initially reported. None of the subsequent studies reported sex differences approaching those found by Horner (1968). Fear of success was higher among females in eight of these studies, whereas males showed more fear of success in the remaining seven. The evidence suggests that fear of success, although potentially an important source of motivation, is not gender specific and appears to be more or less evenly distributed among men and women.

It is also not clear whether the differences found in the original study were due to the sex of the subjects or the sex of the characters. Subsequent studies have found that, when each sex responds to Anne and John, *both* male and female subjects show more fear of success when the central character is a woman (Alper, 1974; Feather & Raphelson, 1974; Monahan, Kuhn, & Shaver, 1974; Robbins & Robbins, 1973). These studies suggest that both men and women are responding according to cultural expectations about the probability of success among men and women rather than a personal desire to achieve or avoid success.

Another problem is the apparent low reliability of the measure. Most studies have reported an interjudge reliability between 80% and 90%, but the percentage of people expressing some fear of success imagery in the studies reviewed by Zuckerman and Wheeler (1975) has ranged from 20% to 88% among women and 9% to 76% among men. Such wide variations across studies suggest that different investigators were using different criteria for scoring protocols. Zuckerman and Wheeler (1975) suggest that the actual reliability of the measure may be as low as .30 to .40. This makes it extremely difficult to interpret findings derived from different studies and may account for many of the inconsistencies found in the literature.

Finally, it is unclear what type of people experience fear of success or how it combines with other traits to determine actual performance. Horner (1968) initially suggested that fear of success occurs more frequently in gifted women with high levels of achievement motivation and found relatively high levels among female honor students. Hoffman (1974) found a similar pattern in both men and women. Kresojevick (1972) and Sorrentino and Short (1974) found a direct relationship between fear of success in women and academic ability, measured by IQ and grade-point average, but other studies have failed to find this relationship. Some studies suggest that fear of success is associated with traditional sex-role attitudes, but research in this area has also failed to show any consistent pattern (Zuckerman & Wheeler, 1975).

Fear of success enjoyed a wave of popularity during the 1970s, but research in the area soon subsided. Methodological problems, low reliability, and inconsistencies in the literature contributed to the general decline. A distinction should be made, however, between fear of success as a theoretical construct and the empirical evidence used to support it. The concept of fear of success has enormous appeal because it helps fill a gap in achievement theory and because it ties in nicely with a relatively large body of data on sex differences and discrimination. Indirect support for the notion that women avoid success because of its negative consequences is provided by three lines of research: (1) research on negative attitudes toward successful women, (2) the portrayal of women in advertising and the mass media, and (3) the internalization of these attitudes by women.

Fear of success, like most attitudes, does not occur in a social vacuum. It is transmitted and maintained to a large extent by attitudes and beliefs common in American society. Sexual discrimination is widespread in America and most women work at low-paying, menial jobs with little prestige and little hope for success. Women are overrepresented in some occupations, such as nursing, social work, and education, and underrepresented in others. Although there are wide cultural differences, male occupations are generally considered more prestigious. Sex differences occur even within occupations. The vast majority of nursery and elementary school teachers, for example, are female, but the proportion of men to women steadily increases from junior high school to high school and on to university. It is as if the really demanding aspects of education are left to men. The overall picture that confronts people in our culture is one of enormous possibilities for men and a more limited range of opportunities for women. Successful women are a minority and often appear out of place in a world designed primarily for men.

Nor is the threat of rejection unrealistic. As women move into more prestigious occupations previously dominated by men, the relative status of those occupations declines (Touhey, 1974). Career women are not always well liked by male work partners (Shaffer & Wegley, 1974) and successful women are sometimes viewed by both sexes as unattractive, immoral, and dissatisfied, and are believed to experience more conflict and difficulty (Monahan, Kuhn, & Shaver, 1974). Progress is occurring, but very, very slowly. A study by Helmreich, Spence, and Gibson (1982) compared attitudes toward women in 1972, 1976, and 1980. Male students showed no overall change between 1976 and 1980 and female students actually showed a small but insignificant shift in the conservative direction. Our culture continues to provide a climate in which successful women expect rejection from both males and females.

Attitudes toward women are also shaped by advertising and the mass media. Several studies (e.g., Friedan, 1963) suggest that the image of women was deliberately manipulated by the advertising industry after World War II to sell products. The strong, self-sufficient woman of the late 1930s and early 1940s gave way to the "beauty queen" and the "happy homemaker." The former was portrayed as shallow and flawlessly attractive, kept eternally young through clothing and cosmetics. The latter was married, devoted to husband and family, and pathologically concerned with cleaning and housework. These were contrasted with the unhappy, conflict-ridden career woman who had given up everything in the selfish pursuit of her own career. The message was simple. You could have a husband and family *or* you could have a career and this message still persists, despite the fact the vast majority of working women are wives and mothers as well.

There is also considerable evidence that women internalize such negative attitudes. Stereotypes about men and women are pervasive. A greater number of desirable traits are assigned to men, and these tend to form a competence cluster (Broverman, Vogel, Broverman, Clarkson, & Rosenkrantz, 1972). Women generally have lower expectations of success (Crandall, 1969; Frieze, 1975), and these differences occur as early as kindergarten. They also occur in school, even though girls receive higher grades. Women approach achievement tasks with less confidence and, therefore, attribute success to unstable factors, such as effort or luck, rather than ability (Deaux, 1976). It is not surprising under these circumstances that many people have come to see success as "unfeminine," and women who score high on measures of status seeking tend to regard themselves as untraditional—that is, different from other women (Ellis & Bentler, 1973).

It is also not surprising that many career women express concern about rejection (Steward & Winter, 1974) or modify their behavior to avoid it. They often consciously lower their level of performance to improve the quality of interaction. These changes tend to be based on their perception of what men think (Hawley, 1971; Horner, 1972), but these perceptions are sometimes wrong (Steinmann & Fox, 1970). Seyfried and Hendrick (1973), for example, found that, whereas women preferred "masculine" men, men were equally attracted to "masculine" and "feminine" women. This discrepancy between what men want and what women think men want helps to underline the notion that fear of success is a relatively stable attitude based on anticipated negative consequences that may outlive the objective social conditions on which the fear is based. If women *believe* success is unfeminine or that it will lead to rejection, they tend to avoid it even though their expectations are wrong.

Research by Travis, Wiley, McKenzie, and Kahn (1983) suggests that men and women also define success differently. When college students were asked to write stories about a time when they had experienced success and failure, women wrote stories that were more socially oriented and concerned with self-improvement. Stories by men described situations that were more task oriented and competitive. Spence and Helmreich (1978) separated achievement motivation into three components—work, mastery, and competition. Whereas college men scored higher than women on mastery and competition, women scored higher on work. A subsequent study by Spence and Helmreich (1983), comparing men and women in various occupations, found that sex differences based on work and mastery had diminished but men were still more competitive. In short, the plausibility of fear of success in women is supported by actual attitudes and practices quite common in our culture, the depiction of women in advertising and the mass media, and the presence of attitudes associating achievement with masculinity. Evidence suggests that such beliefs are acquired early and sustained at least in part by actual social conditions.

More recently, Piedmont (1988) has argued that much of the inconsistency in the literature is due to a misinterpretation of Horner's original theory. He points out that Horner never claimed that fear of success was uniformly present among all women. She claimed that fear of success tended to limit achievement only among women who were high in achievement motivation and then only on tasks that were perceived as being predominantly male oriented. The combined effects of high achievement motivation and high fear of success tends to create an approach–avoidance conflict in which the "desire to succeed is undermined by an anticipation of negative consequences associated with success. As a result, these women compromise their performance in order to maintain affiliation links with others" (p. 467).

Piedmont (1988) suggests that the combination of high fear of success and low achievement motivation may actually enhance performance because of the facilitating effects of arousal. He compared the performance of men and women who were high, mid-range, and low on fear of success and achievement motivation on tasks that were described as masculine, feminine, or neutral and found an interaction on tasks that were described as male oriented. Both men and women performed better if they were high in achievement motivation and low in fear of success or low in achievement motivation and high in fear of success. They performed poorly when their levels of achievement motivation and fear of success were either both high or both low.

Piedmont's (1988) results seriously challenge much of the research that suggests fear of success is not a useful construct. Fear of success must be considered within the context of other motives, such as the motive to succeed and the fear of failure. It can be expected to have a debilitating effect only on a small subset of people who are simultaneously high in both achievement motivation and fear of success and then only on tasks that are perceived as predominantly masculine. The failure to find significant differences in much of the previous research is due (in part) to the assumption that all women experience fear of success and that it can be studied without considering the level of achievement motivation. This creates a "sampling bias through the overinclusion of many false positives (females believed to be high on FOS, but who are not)" (p. 469).

There is little evidence, however, to support the view that fear of success is limited to women. Empirical studies have found few sex differences and there are general trends within our society that suggest men may experience fear of success as well. One could speculate that fear of success would be common among men in *any* group where low achievement is the norm (see Chapter 5). Success in such groups would bring the stigma of appearing odd or pretentious and the possibility of social rejection. The pattern may be quite different for men and women. Moore (1972), for instance, found that high levels of education and occupational status were related to fear of success in white women, but among men fear of success was associated with low status and less education. There is also evidence that many blacks in the United States associate hard work and success with "acting white" and a kind of going over to the other side (Ogbu, 1986; Petroni, 1970). Although the antecedents are different, the effects are much the same. Many people in our society avoid success because it is associated (consciously or unconsciously) with unwanted traits and social rejection.

ATTRIBUTION MODELS OF ACHIEVEMENT MOTIVATION

During the 1960s and early 1970s, various attribution theories were developed in an attempt to explain how people perceive and understand the *causes* of behavior (e.g., Heider, 1958; Jones & Davis, 1965; Kelley, 1967, 1972). These included models that attempted to explain previous motivational constructs, such as achievement motivation and fear of failure, in attributional terms. Weiner et al. (1972), for example, proposed a two-dimensional model based on the locus of causality and the amount of stability. These two dimensions lead to four distinct attributions, as shown in Table 3.1.

Although the original theories assumed that people process information accurately, it soon became apparent that errors in person perception are quite common. Most people distort their self-perception in a way that enhances self-esteem, but others take too little credit for success and too much responsibility for failure. There are also very pronounced *gender differences* in the way men and women explain success and failure, particularly when the tasks are perceived as masculine. Males generally attribute their success

TABLE 3.1. Two-Dimensional Attribution Model for Success and Failure

	Stable	Unstable
Internal	Ability	Effort
External	Task difficulty	Luck

to ability and their failure to a lack of effort, whereas females tend to attribute their success to effort, luck, or an easy task and their failure to a lack of ability (e.g., Dweck & Reppucci, 1973; Nicholls, 1975). When men and women perform equally well, women are seen as trying harder (Feldman-Summers & Kiesler, 1974; Taynor & Deaux, 1973). On male-oriented tasks, women's success is attributed more to an easy task or luck than to ability (Deaux & Emswiller, 1974; Feather & Simon, 1975). Stable attributions for failure among women are also more likely to generalize to new tasks and evaluators and from one school year to another (Dweck, Goetz, & Strauss, 1980). For extremely high levels of performance, men and women are evaluated more or less the same because both effort and ability are seen as necessary but, for lower levels of performance, men receive more credit than women—even when the level of performance is precisely the same.

Gender differences in attributions for success and failure are partly due to initial *expectations* about the probability of success among males and females. Kay Deaux (1976) has noted that both men and women consider men more competent and therefore expect them to perform well. When men succeed, these expectations are confirmed and observers attribute the success to ability. Success among females, on the other hand, is unexpected and this leads observers to seek unstable attributions to explain it. The same argument applies to failure. For men, failure is unexpected and is therefore attributed to unstable factors, such as bad luck or lack of effort. For women, failure is expected and is therefore attributed to stable, internal factors, such as lack of ability.

Gender differences also reflect the way boys and girls are treated by adults. Carol Dweck (1975) argues that the more favorable treatment received by girls tends to promote a state of academic helplessness. Boys tend to be criticized for a variety of reasons—for their behavior, the way they dress, their lack of motivation, as well as poor school performance. They therefore learn to disregard negative feedback because it is so common. Girls, on the other hand, are usually neater and better behaved, and it is often assumed that they are trying as hard as possible so their criticism is limited to academic performance. They take criticism more seriously and often give up at the first sign of negative feedback.[4]

Irvine (1985, 1986) has also found that girls receive less attention from (female) teachers *in general.* She and her associates have observed well over a thousand student–teacher interactions in 67 different schools and have found that male students are more likely to initiate interactions by talking

[4]Although girls generally respond in a more helpless manner to teachers and other adults, Dweck and Bush (1976) have found a similar pattern of response among boys to peer evaluators. They suggest that, for boys, the conflict between adult and peer values causes boys to violate adult standards to be accepted, and this makes them relatively insensitive to adult feedback but more sensitive to similar evaluations from peers. These differences will be discussed in greater detail in Chapter 5.

out loud, answering questions without being recognized, and dominating class discussions. Black girls in lower elementary grades are more active and receive more attention than white girls, but by late elementary school, they have been socialized into what appears to be the traditional female role. Brophy and Evertson (1981) have found that even when female students initiate contact with teachers they are frequently ignored. Irvine points out how ironic it is that girls who have been socialized to be more on-task and less disruptive receive less attention and this tends to reinforce the belief that boys are more important.

Dweck and her associates have identified two different patterns of achievement-related behaviors, which they describe as *mastery-oriented* and *helpless* (e.g., Diener & Dweck, 1978; Dweck, 1975). Mastery-oriented students are able to sustain high levels of motivation even in the face of failure. They are not easily discouraged by minor setbacks and generally regard failures as temporary and surmountable. They enjoy working on different tasks, are able to concentrate fully on whatever they are doing, and generally improve their performance over time. A second group displays a "learned helpless" pattern. They find new tasks aversive, have difficulty sustaining motivation or interest on difficult tasks and are quickly discouraged by minor setbacks. They also find it difficult to concentrate, and their thoughts frequently wander to task-irrelevant areas. The helpless pattern is characterized by avoidance of challenge and deteriorating performance in the face of failure. It also seems to be much more common among girls.

Diener and Dweck (1978) have studied the kinds of attributions these two groups make by having subjects verbalize their thoughts during problem-solving. Boys and girls in the fifth grade were divided into helpless and mastery-oriented groups on the basis of their response to items from the Intellectual Achievement Responsibility Scale (Crandall, Katkovsky, & Crandall, 1965), which is essentially a measure of locus of control. There were few differences between mastery-oriented and helpless children when they succeeded but, during failure, helpless children were far more likely to attribute their failure to a lack of ability. They also begin to express more negative feelings about the task and engage in more task-irrelevant statements. Some subjects expressed clear signs of resignation by saying such things as "I give up" or "Nothing I do matters." Mastery-oriented students, on the other hand, made few explicit attributions. Their verbalizations consisted primarily of self-monitoring and self-instructions aimed specifically at improving performance. Some even expressed delight that they would now be able to test their ability. When mastery-oriented children did make attributions, they tended to attribute their success to ability and their failure to unstable factors, such as bad luck, lack of effort, or the increased difficulty of the tasks. Helpless and mastery-oriented students also differed substantially in their subsequent level of performance. In striking contrast to helpless children who showed a progressive drop in their level of problem solving, 80% of mastery-oriented children either maintained or improved their level of performance when provided with negative feedback. The attributions made by helpless children were not the result of a careful causal analysis but seem to be a more or less "automatic" response to failure.

Dweck (1975) has also found that helpless children can be *retrained* to change their typical attribution pattern. She assigned helpless children to two different conditions. One group was taught to take responsibility for their failure and attribute it to a lack of effort rather than a lack of ability. A second group was provided with a series of tasks at which they always succeeded. The behavior of children in the attribution-retraining condition changed dramatically. Not only did they change the perceived cause of behavior but failure began to lead to improved performance. Those in the "success only" condition showed no change in performance on subsequent trials. Some limitations to attribution-retraining research will be discussed in Chapter 11.

The helpless response pattern can be seen as maladaptive because obstacles and challenge are an inherent part of most important activities and any response that discourages people from seeking challenge or dealing with it effectively tends to limit the development of new skills. The helpless and mastery-oriented patterns can be seen as two distinct reactions with strikingly different emotional, cognitive, and behavioral consequences. Although these patterns were first found in children under laboratory conditions, subsequent research has found similar patterns in adults and among children in real-life situations as well.

Dweck (1986; Dweck & Elliot, 1983) has recently extended her model by making a distinction between two types of goals—*learning goals* and *performance goals*— that roughly correspond to an intrinsic desire to learn and a desire to perform well. Children with learning goals believe that intelligence can be improved and strive to increase their level of competence, develop their ability, and master new material. They tend to be mastery oriented and show the same type of traits described previously. Children with performance goals believe that intelligence is fixed and tend to focus on avoiding criticism or obtaining positive evaluations of their own competence. They may be either mastery oriented or helpless but a mastery orientation requires a high level of perceived ability.

Children with learning and performance goals do not differ in terms of their initial level of ability, but these patterns can have a profound effect on cognitive performance. According to Dweck and Leggett (1988), some of the brightest and most skilled individuals tend to adopt the maladaptive helpless pattern, and somewhat ironically, bright children who are most concerned with their own level of performance behave in a way which is counterproductive. Even when those with performance goals have high levels of perceived ability, they may sacrifice learning opportunities that include the possibility of failure for a chance to look smart (Elliott & Dweck, 1985). For children with performance goals, the main goal is to impress others and satisfaction is based on performing well. For these children, effort and ability are seen as inversely related because high effort and failure suggest low levels of ability. In the face of uncertainty, children with performance goals reduce effort, and the fear of failure may reduce or eliminate any intrinsic interest in the task itself (Elliot & Dweck, 1985). Children with learning goals, in contrast, derive satisfaction from trying hard and see effort as a means for using their talents, overcoming obstacles, and increasing their level of ability.

Dweck and Leggett (1988) have subsequently made a more explicit attempt to link learning and performance goals to *implicit theories of personality.* They point out that people with performance goals tend to view intelligence as a fixed entity that does not change substantially over time, whereas those with learning goals tend to view it as a collection of specific skills that can be developed and improved through challenging activities. They have identified five distinct deficits associated with a performance-goal approach. When people with performance goals encounter difficulties, they frequently experience a feeling of inadequacy that suggests success may not be possible. They then tend to defensively reduce effort in order to maintain their sense of competence because high effort and failure are associated with a low level of ability. Their concern about failure and the negative evaluation it implies causes them to divide their attention between themselves and the task making it difficult to concentrate. The task tends to be devalued, and this prevents them from achieving any intrinsic satisfaction from its solution. Those with learning goals, in contrast, tend to believe in their own ability, confront the task as an opportunity to develop new skills, and increase effort. The task actually becomes more attractive, and this increased attraction to the task enhances concentration and intrinsic enjoyment.

Dweck (1986) argues that the current approach to teaching, which uses frequent praise for small improvements in performance sets up a performance-goal orientation. Children come to expect praise and seek positive evaluations rather than focusing on the task itself. Children should learn to deal with both success *and failure* by attributing success to ability and failure to unstable factors, such as lack of strategy or effort. Finally, Dweck and Leggett (1988) have pointed out that children from lower socioeconomic classes consistently tend to adopt performance goals in which their concern centers around avoiding criticism by avoiding challenge altogether.

Variations in achievement motivation and the motive to avoid success and failure show how naive it is to assume that people try as hard as possible and that differences in performance are based on actual ability. Many people do not try, either because they are unmotivated or because they fear the negative consequences associated with failure. Others actively strive not to succeed because they fear rejection and social disapproval. It may be that only a tiny fraction of the general population acquire high levels of achievement motivation uncontaminated by the conflicting drives to avoid success or failure. Everyone else operates well below his or her full potential.

CHAPTER 4

Internal-External Locus of Control

People's perception that they can control the events in their day-to-day lives has a profound effect on the way they seek out, perceive, and process information. This chapter explores the general concept of locus of control, its childhood antecedents, and some of the sex and racial differences that have been found. There are profound differences in the way blacks and whites in the United States develop a sense of locus of control and the social conditions of blacks are explored in detail. The chapter ends with a brief overview of the related concept of "learned helplessness" and a discussion of the small group of resilient people who have overcome adversity and been strengthened by their negative childhood experiences.

During the 1970s, there was a shift in emphasis from achievement motivation as an acquired drive to cognitive factors underlying goal-directed behavior. This shift in emphasis drew from two different sources. The first was the growing research on attribution models derived in part from the theoretical work of Fritz Heider. The second stemmed from Julian Rotter's social learning theory and, more specifically, his concept of locus of control. Rotter's 1966 monograph on locus of control was the most cited article in the social science literature between 1969 and 1977, being cited more than twice as often as any other work (Garfield, 1978). There are thousands of studies on locus of control in virtually all areas of psychology, and locus of control scales have been translated into many foreign languages where they have become the source of much additional research. Although only a fraction of the research on locus of control focuses on social origins of mental ability, that fraction is still sufficiently large to warrant a separate treatment.

Rotter's concept of locus of control, however, is only one of many theories concerned with perceived control. Rotter's (1966) original concept was linked to other theories, such as alienation (Merton, 1946), competence (White, 1959), autonomy (Angyal, 1941), and achievement motivation. Since then, a number of related concepts have been developed, such as self-efficacy (Bandura, 1977, 1986), personal causation (deCharms, 1968, 1981), and learned helplessness (Seligman, 1975). Recent experimental research has shown that the perception of control is a major factor influencing many aspects of behavior. The literature on perceived control is far too large to review briefly but some examples may be useful.

Glass and Singer (1972) and their associates have carried out a series of experiments designed to assess the effects of loud noise on a variety of tasks. They found that uncontrollable loud noise interferes with proofreading and problem solving. When subjects were told that they could turn the noise off by flipping a switch (but were encouraged not to), the decrements disappeared and they attempted many more unsolvable puzzles and made significantly fewer errors. In a related study, the ratings of painful electric shock and their aftereffects were reduced when subjects believed that they could control the duration of the shock (Glass et al., 1973). These studies show that it is not the aversive stimuli *per se* which determines behavior but the combined effect of negative stimuli and the perceived lack of control. The negative effects of noise and electric shock are considerably reduced when people believe that they can control them, even though the stimuli are precisely the same.

Dramatic effects have been found in real-life situations as well. Schultz (1976) studied the effects of control on elderly residents in a nursing home by dividing them into four groups. Volunteers visited one group at times selected by the residents. A second group was notified in advance when they would be visited. A third group received visitors unexpectedly for an equivalent amount of time, and a final group was not visited at all. Schultz found that the first two groups were subsequently rated as healthier, required less medication, and were more active, hopeful, and happy. They also felt less lonely and bored than the groups who were not foretold of the visits and those who were not visited at all. Ironically, those in the groups with foreknowledge and control deteriorated rapidly and showed significantly poorer health after the visits came to an end.

A similar study by Langer and Rodin (1976) divided residents in a nursing home into two groups and manipulated the amount of control. Residents on one floor were given a speech that stressed personal responsibility and were allowed to select and take care of a plant. A second group received a message that stressed the staff's responsibility for them and had a plant placed in their room that was cared for by the staff. Members of the first group were significantly more happy, active, and alert. They also spent more time interacting and talking with others. In a follow-up study, Rodin and Langer (1977) found the effects of personal control were still present 18 months later. Even more striking were dramatic differences in death rates. During the intervening period, 15% of those in the first group died compared with 30% of those in the group without personal control.

These are just a few of the many studies showing that the perception of control can play a powerful role in our day-to-day lives. The effects occur even when there are no real differences in stimulation or objective circumstances.

It is the *perception* of control, not necessarily the actual control, that reduces the negative effects of aversive stimulation and leads to improved health among the elderly. Although these studies have manipulated the perception of control, Rotter's theory treats locus of control as a relatively stable personality variable that develops quite early, is sustained by concrete social conditions, and affects a wide range of behaviors. This chapter will focus on the social origins of locus of control both within the family and society at large and the cognitive consequences of perceived control. A final section is devoted to the related concept of learned helplessness.

ROTTER'S BASIC THEORY

Julian Rotter's (1966) concept of locus of control is simply one aspect of his more general *social learning theory.* According to Rotter (1954), three factors determine behavior—reinforcement, expectancy, and the psychological situation. *Reinforcement* is based on the importance or desirability of a particular goal. It varies from person to person according to the previous history of reinforcement, so that the same outcome may be highly valued by one person but not by another. Individuals tend to be relatively consistent in the value they place on different goals, and these preferences transfer from situation to situation. The *situation* in Rotter's theory refers to the psychological context—the situation from the perspective of a particular person. *Expectancy* is based on a person's estimate that a particular outcome can be achieved. It too is based on previous experience with both similar tasks and tasks in general. People who have performed consistently well on particular tasks can expect to do well in the future. The expectancy is a *belief* about the possibility of reinforcement, but it need not be accurate. A person may expect to do well and fail or may expect to do poorly on tasks which could be easily accomplished.

Social learning theory predicts that a person who perceives two situations as similar will generalize the expectancy of obtaining reinforcement from one situation to another. Repeated experience with similar situations produces a specific expectancy in a particular area, such as academic achievement, sports, or social approval. But there is also some generalization across dissimilar tasks. There are thus two kinds of expectancies. One involves specific expectancies for a particular type of task, whereas the other is a *generalized expectancy* based on the overall past history, sequence, and pattern of reinforcement. If a girl, for example, has mastered the game of chess, she can expect to do reasonably well when she plays again, but she may also expect to excel at other games, such as bridge, Monopoly, or Trivial Pursuit. Specific expectancies are important for predicting performance on familiar tasks, whereas the relative importance of generalized expectancy is greater in ambiguous or novel situations.

Internal-external *locus of control* is simply a measure of this generalized expectancy. It is a continuum that ranges from extreme internality at one end to extreme externality at the other. Those who are internal see reinforcement as contingent on their own behavior. They assume that skill, hard work, foresight, and responsibility will pay off, whereas those who are external believe that what happens to them occurs as a result of forces beyond their control, such as luck, powerful others, or political institution. They are more reluctant to initiate behavior because they do not believe that it will make a difference. While many people have simply assumed that locus of control is the central concept within social learning theory, it is not (Rotter, 1975). Locus of control is just one of many factors that determine behavior in a particular situation. Others include reinforcement value, specific expectancies, and the situation itself. For this reason, Rotter predicts a low but sometimes significant correlation between locus of control and a wide range of behaviors.

Rotter's (1966) measure of locus of control consists of 23 items related to perceived control plus 6 filler items used to disguise the purpose of the test. Each item consists of a pair of alternates and people are asked to select the statement in each pair that they most strongly believe. The following items are examples (with the external choice underlined):

1. A. Many of the unhappy things in people's lives are partly due to bad luck.
   B. People's misfortunes result from the mistakes they make.
2. A. Becoming a success is a matter of hard work, luck has little or nothing to do with it.
   B. Getting a good job depends mainly on being in the right place at the right time.
3. A. With enough effort we can wipe out political corruption.
   B. It is difficult for people to have much control over the things politicians do in office.

The overall score is based on the total number of external items selected and ranges from 0 to 23.

Locus of control is a continuous distribution with two extremes. Most people score near the center on the scale. When the scale was first developed, the average score was near 8, but during the 1970s, the mean shifted to between 10 and 12, suggesting a general trend toward externality for the population as a whole. Much of the shift can be accounted for by a small subset of items that measure the perception of political control and seems to suggest a growing disenchantment with the political system. Locus of control is not a typology. People can be described as more or less internal or external and a median split is frequently used to assign people to groups. Those in the external group score relatively higher than those in the internal group and can be assumed to possess more external traits. Locus of control is useful for comparing groups within a study, but Rotter (1975) warns against using it to make individual predictions because the correlation with behavior is typically quite small.[1]

At about the same time that Rotter's (1966) monograph appeared, Crandall, Katkovsky, and Crandall (1965) introduced a similar measure, the Intellectual Achievement Responsibility Questionnaire, which focuses on children's success and failure in school. Battle and Rotter (1963) and Nowicki and Strickland (1973) have also developed tests for children, and Nowicki and Duke (1974) and Mischel, Zeiss, and Zeiss (1974) have devised measures for those in preschool. Finally, several forms have been constructed to measure internal and external beliefs among the elderly (Nowicki & Duke, 1983; Reid & Ziegler, 1981) and it is now possible to make age-related comparisons virtually from cradle to grave.

SOCIAL SOURCES

Rotter's theory suggests that an internal locus of control should be associated with a warm and consistent home environment. Internal expectations are based on a perceived link between reinforcement and behavior, and such a link can only occur when children are allowed to exert some control over things that happen to them. Once expectations develop, they then help determine the subsequent interpretation of similar events and, to a lesser extent, events in general. Although it is widely assumed that warmth and consistency are associated with the development of an internal perspective, research on childhood antecedents has produced a far more complex picture.

Katkovksy, Crandall, and Good (1967) found that children with an internal locus of control were more likely to have mothers who were nurturant, affectionate, and protective and who offered praise and approval instead of criticism. Internals describe their parents as less withdrawn and rejecting, less likely to use hostile control, and more positively involved and consistent in their use of discipline (Davis & Phares, 1969). Shore (1967) found that male externals perceived their parents as using more psychological control and as being less warm and intrinsically accepting.

Although the overall pattern seems to suggest some relationship between locus of control and a warm, supportive home environment, there

[1] Although locus of control has had a fairly good track record, it has not been without its critics. The most common complaint is that it is not a unified concept but consists of a number of different factors (e.g., Collins, 1974; Levenson, 1973, 1981; Mirels, 1970). Others point out that the amount of perceived control may vary with the activity. Some people may feel that they can do well in school, maintain their health, or handle themselves in social situations, and yet feel totally unable to alter larger social or political institutions. This has led to the development of more limited scales that focus on particular problems and domains.

are significant *gender differences.* Katkovsky, Crandall, and Good (1967) found a positive relationship between internality and maternal support for boys but no significant pattern for girls. Remanis (1971) found that women who felt rejected by their mothers tended to be more *internal.* He reasoned that a warm, protective home may encourage female dependence, whereas girls who feel rejected may be forced to become independent to satisfy their needs. Levenson (1973) used her own scale, which distinguishes among beliefs in internal control, chance, and powerful others, and found that boys who were helped and tutored by their mothers tended to be more internal, but girls who described their mothers as protective tended to be external. Sex differences are not limited to American society, however. McGinnies, et al. (1974) compared locus of control scores among students in Australia, Japan, New Zealand, Sweden, and the United States and found that girls were more external in each case.

To further murky the waters, Crandall (1973) found a significant relationship between internality and maternal "coolness" and "criticality" among teenagers and young adults. She suggests that, whereas warmth and protectiveness may be necessary during childhood, some degree of coolness, criticalness, and stress is important later to prevent dependence. Mothers of internals rewarded dependence less often, showed less involvement, and were more likely to have "pushed their children from the nest."

The overall pattern, at least for boys, seems to be one in which mothers provide a warm, stable, and protective home during early childhood yet encourage independence as well. The parents provide reasonably high standards, offer suggestions rather than directives, and use praise and approval rather than criticism. The stress on independence increases as the child matures so that internal adolescents frequently receive a "push from the nest" that prevents overdependence and fosters a belief in personal control. Externals are more likely to be raised in hostile, unpredictable, or authoritarian homes where discipline is stressed and independence is discouraged. The pattern for girls is less consistent. Internality sometimes reflects a similar pattern in both sexes, but some studies have found a direct relationship between externality and paternal protection among girls. Because girls are typically more external than boys, additional factors may limit their perception of control.

Locus of control is also altered by *early childhood trauma,* particularly during the preschool years (e.g., Bryant & Trockel, 1976; Nowicki, 1978). Duke and Lancaster (1976) found that children from father-absent homes were significantly more external than those from intact homes and speculate that the loss of a father was equivalent to a "massive dose of fate." Hetherington (1972) found that girls without fathers were more external, particularly when the loss occurred early in their lives. Marine recruits from intact homes have also been found to be more internal than those from homes where divorce or separation had occurred (Cook, Novaco, & Sarason, 1980, cited in Lefcourt, 1982).

There are also *social-class differences* in locus of control. Lower-class individuals tend to be more external than those from the middle class. Stephens and Delys (1973) found that not only were those from the lower class significantly more external than those from the middle class but children from lower-class homes above the poverty line were less external than those below. Although research on social-class differences tends to be correlational and cannot really explain the origins of these differences, they probably stem from a blend of factors occurring within the home and society at large. Lower-class parents are typically more authoritarian and stress obedience and respect for authority, whereas middle-class parents, particularly professionals and those who are self-employed, place more emphasis on independence and are more likely to use suggestions rather than directives. But parents also serve as models for their children, and those in the lower-class typically have less access to power and little control over their own lives. As Phares (1973) has pointed out, those groups "which have little access to real power or material advantages, and whose members perceive their overall movement within society as being greatly limited, will more likely be external in their belief systems" (p. 18).

Finally, there are *racial differences* in locus of control. One of the most consistent findings in the locus of control research is that blacks are more external than whites. In fact, it is frequently difficult to tease out the separate effects of class and race because blacks are so disproportionately represented among the poor. There are also substantial differences in the social environments of poor blacks and poor whites that will be discussed shortly. Battle and Rotter (1963) found that lower-class blacks were significantly more external than either middle-class blacks or lower-class whites. Race and social class thus appear to interact and enhance the perceived lack of control. Not all studies have found significant racial differences, but as Lefcourt (1982) has pointed out, despite the occasional failure, whenever differences are found, blacks are always more external than whites. Coleman and his associates reported in their landmark Equality of Educational Opportunity Study that perceived control accounted for more variance in black school achievement than any other variable, including school, teacher, or family background (Coleman et al., 1966).

While it is possible that child-rearing plays some role in the development of an external perspective among blacks, several lines of research suggest that externality is due more to the sociocultural context. First, observational studies of black children in the home have generally shown that they receive a great deal of warmth and affection but are also encouraged to become independent and self-reliant at a relatively early age (e.g., Ogbu, 1985). These are precisely the conditions that promote an *internal* perspective among white boys. Second, unlike whites who tend to become more internal as they grow up, externality among blacks tends to increase with age (Duke & Lewis,

1979). There also appears to be a direct relationship between externality and intelligence among lower-class blacks. More intelligent black children from lower-class homes tend to be more external than those with lower levels of ability (Battle & Rotter, 1963).

Finally, there is some evidence that blacks make a clearer distinction between personal control and general or ideological control. Lao (1970) found that an internal belief in personal control is positively related to academic performance, academic confidence, and level of aspiration in predominantly lower-class black college students, whereas a tendency to blame the system is associated with civil rights participation and collective social action. She speculated that this distinction is less salient among whites and the two factors tend to occur together, thus creating a unified dimension. These studies suggest that an external orientation among blacks is based on a growing awareness that lower-class blacks have relatively little control over the economic and political institutions that shape their lives. To understand these beliefs, it is necessary to pause briefly and examine the concrete social conditions of blacks in the United States today.

BLACKS IN AMERICAN SOCIETY

Despite the common belief that blacks have made tremendous gains since the civil rights movement in the 1960s, the actual picture is far more complex. There have been two striking developments since the 1960s. The first is the emergence of a new *black middle class*—better educated, housed, and paid than ever before. The proportion of blacks earning over $25,000 (in 1982 dollars) increased from 10% in 1960 to almost 25% in 1982, and the number of black professionals, technicians, managers, and administrators increased by 57% between 1973 and 1982 (Wilson, 1987). Blacks now represent 7.4% of accountants and auditors, 9.4% of all teachers, and 18.1% of social workers (Gelman, Springen, Brailsford, & Miller, 1988). By 1987, there were almost 800,000 black managers and executives—5.6% of the total, compared with 500,000 four years earlier—and close to 7,000 elected black officials (Kantrowitz & Springen, 1988).

Closely related to the first trend is the gradual development of a new group of truly disadvantaged blacks who are cut off from the rest of society and have little hope for immediate improvement. As more wealthy and better educated middle-class blacks move from the ghetto to take advantage of new opportunities, they have left behind a hard core of socially disabled people, sometimes referred to as the "*underclass.*" The underclass is defined not so much by poverty as by behavior and the statistics provided by the Urban Institute (cited in Alter, Brailsford, & Springen, 1988) are staggering. Nearly half the people in the underclass live below the poverty line, over a third of the households are on welfare, and 60% are headed by women. Thirty-six percent of teenagers are high school dropouts, and 56% of men are not regularly employed.[2]

The deteriorating conditions of lower-class blacks can be traced to *structural changes* within the black community and within the nation as a whole. The rapid economic expansion following World War II drew blacks from the rural South to industrial centers in the North. Black-owned newspapers, such as the *Chicago Defender,* described the North as a "promised land" (Slaughter & McWorter, 1985), offering high salaries and full employment. The shift from blue-collar to white-collar jobs during the postindustrial period eliminated many positions within the inner city created during the postwar economic boom and produced a rapid increase in poverty and unemployment. The loss of jobs in the manufacturing sector was more than offset by an increase in the service industries, but virtually all the new jobs occurred in the suburbs and nonmetropolitan areas, far removed from the growing concentration of urban poor. Much of the movement by middle-class blacks out of the ghetto is based on the desire to go where opportunities occur. But for those left behind, high unemployment and a lack of opportunity is often the norm. The proportion of black men who are regularly employed dropped from 80% in 1930 to 56% in 1983 and the rate of unemployment is particularly high among young blacks in the inner city.

Wilson (1987) has argued that the high frequency of unemployment among black youth has led to a decline in the number of "marriageable" men and is directly responsible for the drastic increase in illegitimate births and single-parent homes. More than half of all black children are now born out of wedlock and roughly 40% of these have teenage mothers. In 1940, 72% of black families with children under 18 years were headed by men. By 1983, nearly half of all black families were headed by women. Black women are far more likely to delay marriage, separate without divorce, and not remarry if widowed or divorced. This has led to a drastic increase in children growing up in poverty because the median income of female-headed families in 1983 was only 43% of that of married couples.

The increase in illegitimate births has gone hand in hand with *changes within the home* itself. Previously, single mothers moved in with their own mothers or other relatives, and single-parent families were incorporated within extended families headed by older adults. This network of kinship often included as many as four or five generations, welded together by a sense of mutual obligation and generally guided by one dominant person. These

[2] The stress on these two groups is not meant to suggest that blacks can be divided into a nonmetropolitan middle class and an urban poor. A substantial number of poor blacks still live in rural areas, and an even greater number of working-class blacks live in urban centers. Changes within these groups have undoubtedly occurred as well, but they are less dramatic than those occurring within the black middle class and among the urban poor. Changes among underclass blacks are particularly important because of the growing number of black children living below the poverty line.

families provided mutual support during hard times, shared resources, and generally helped each other out. Between 1969 and 1973, 60% of all black children born out of wedlock were raised in extended families. Between 1974 and 1979, the proportion had dropped to 37% (Wilson, 1987).

Although remnants of the extended family may persist among households that are physically separated, children brought up by single parents experience hardships not common among those raised in nuclear families or extended homes. Such children are far more likely not only to experience poverty but to be persistently poor for long periods of time. Nearly half of all black children live below the poverty line, and 75% of these are living in families headed by women. Mothers in such homes experience more social isolation, often raise their children without help, and are more prone to psychological stress. Young black women with illegitimate children are further handicapped by the disruption of school and by the fact that they virtually never receive child support. They are also far more likely to have daughters who become unwed mothers, thus perpetuating a vicious cycle of poverty which passes from generation to generation.

Despite a high rate of poverty within ghetto neighborhoods throughout the first half of the century, social disruption did not reach catastrophic proportions until the 1970s. Previously, inner cities included a "vertical integration," in which lower-, middle-, and working-class families lived together, shopped together, and sent their children to the same schools (Wilson, 1987). The very presence of middle-class families provided role models which helped promote the perception that education was worthwhile, that employment was a viable alternative to welfare, and that family stability was the norm. Middle-class families served as a *social buffer* that helped provide stability and absorbed the shock of uneven economic growth. The increased exodus of middle-class families makes it difficult to sustain basic institutions such as churches, recreational facilities, and schools. As basic institutions decline so does the sense of community spirit, neighborhood identification, and social sanctions that discourage aberrant behavior and crime. The inner city becomes a less pleasant and more dangerous place to live. Deterioration beyond a certain point seems to set off an exponential increase in the rate of social disintegration and violent crime.

As poverty and unemployment increase, welfare, crime, and underground activities become not only viable alternatives but a way of life. About 60% of the children born out of wedlock and not adopted receive welfare. Only one in nine Americans is black, but in 1984, nearly half of all those arrested for murder and nonnegligent manslaughter were black. Blacks also make up 61% of those arrested for robbery and 38% of those incarcerated for aggravated assault. Forty-one percent of all murder victims are also black, and homicide is the leading cause of death among black men and women aged 25 to 34 (Wilson, 1987).

There are also some fundamental differences between black and white poverty. Only 7% of all poor whites live in extreme poverty areas (where the rate of poverty is above 40%), but 39% of blacks live in extreme poverty areas. Poor blacks and whites, therefore, typically inhabit areas that are ecologically and economically very different. Poor whites are far more likely to live in neighborhoods with a substantial number of middle-class families, and they have access to better quality shopping areas, recreational facilities, and schools (Wilson, 1987).

Wilson (1987) has pointed out that the deteriorating conditions among poor blacks cannot be attributed to racism per se. There is no reason to assume that racism is more rampant in the 1970s and 1980s than it was during the 1940s, 1950s, and 1960s. The changes are due to structural changes within the economy that have led to unemployment and indirectly to increases in illegitimate births, violent crime, and single-parent homes. This means that traditional social psychological solutions, such as changing attitudes, will no longer have much effect. Simply changing the attitudes of white Americans cannot improve conditions, although it may help to some extent. He goes on to suggest, "There is an economic structure of racism that will persist even if every white who hates blacks goes through a total conversion" (p. 11).

The deteriorating conditions among poor blacks within the inner city have been exacerbated by other changes occurring within American society at the same time. These include the disintegration of the civil rights movement, withdrawal of federal support, and a decline in media coverage. During the 1960s, although conditions were bad, there was a general feeling that conditions would improve and a belief that individuals could help change the world by becoming personally involved. The assassination of many of the major civil-rights supporters—John Kennedy, Robert Kennedy, Malcolm X, and Martin Luther King—shattered these beliefs, leaving the movement somewhat rudderless and adrift. The old civil-rights network made up of students, academics, community-service workers, and government officials fell apart, and the former optimism has been replaced by resignation.

A related development is the *withdrawal of federal support.* In 1964, Lyndon Johnson declared a "war on poverty," aimed in part at the black community. This war on poverty went through a period of benign neglect during the Nixon administration and something less benign under Reagan and Bush. The Justice Department, which once championed black causes, has turned its attention to fighting such things as busing and affirmative action and has become increasingly adversarial toward blacks in court cases.The effects of the Reagan budget cuts have been studied by both Joe (1982) and Moscovite and Craig (1983), who have found that millions of partially dependent families were either removed from welfare or suffered severe declines in their living standards. This neglect has allowed problems to fester and grow, but it has also altered the perception of the federal government. A *Newsweek* poll conducted in 1988 showed that 71% of blacks (but only 29% of whites) felt the federal government is doing too little to help black Americans.[3]

A final factor is the role of the *mass media.* On the one hand, television has increasingly used black entertainers, creating the impression that the civil rights movement was a success. On the other, there has been a steady decline in the coverage of black issues. Naisbitt (1984) has done a line-by-line analysis of the amount of newspaper space devoted to various topics and has uncovered a disturbing trend. In the late 1960s and early 1970s, civil rights was the most frequently covered topic and the coverage focused almost exclusively on black civil rights. By 1973, concern with the environment became—on a line-by-line basis—the most covered issue. Beginning in 1969, there was a growing concern with sexual discrimination, and by 1975, material on sexual and racial discrimination appeared equally often. Starting in 1977, both subjects begin to give way to concerns about age discrimination until two-thirds of the space devoted to discrimination focused on ageism, with sexism and racism sharing the remaining third.

These shifting trends do not mean that issues are being resolved. Between 1970 and 1979, the amount of space devoted to drug use dropped 92%, although the use of hard drugs actually increased. Coverage simply reflects what reporters and editors consider newsworthy. It is quite easy for white Americans to conclude that racial problems have been resolved, and this is supported by the visible gains made by the black middle class and by black entertainers on television. Underclass blacks, on the other hand, who know better may conclude (perhaps correctly) that the majority of whites in American society simply no longer care about civil rights or are unwilling to invest the time and energy necessary to make things better.

The combined effects of black flight from ghetto neighborhoods, disintegration of the civil rights movement, government neglect, and reduced media coverage lead many blacks to feel that white Americans simply do not care about blacks, except as a source of crime. This can be seen in the same *Newsweek* poll cited earlier. White Americans were far more likely to believe that conditions among blacks have improved and to oppose affirmative action, whereas blacks more frequently believe that white people want to keep blacks down, that middle-class blacks are not doing as much as they should to help improve conditions for poorer blacks and that blacks are treated more harshly by the criminal justice system. Surprisingly, both blacks and whites agreed about the cause of poverty. Over 40% of both groups believed that the fault lies within the system, whereas 30% of each group blamed blacks themselves.

[3]The federal government's attitudes toward blacks can also be seen in more concrete terms, such as hiring practices. The United States Supreme Court, which is supposed to be a bastion of social justice, has hired only two black law clerks during its entire history. In 1979, there were 32 clerks—27 men and 5 women—all white. In contrast, each justice of the court has a messenger who runs errands. All of these, except one who came from Thailand, were black (Pinkney, 1984).

In short, there is ample reason to believe that an external locus of control among blacks stems from their realistic perception of social conditions in American society. Among those in the underclass, a belief in an external perspective is "overdetermined." The effects of poverty and father-absenteeism combine with the lack of successful role models to produce a picture of a society out of control. The frustration and stress often lead to random and seemingly inexplicable acts of aggression on members of the black community itself. Blacks now account for half of all violent crimes, and the most frequent perpetrators and victims are young black men (Wilson, 1987).

Many people who acknowledge the conditions of blacks in American society still believe that blacks are responsible. There is an ongoing debate between liberals and conservatives about who is to blame. Any simple-minded, either-or approach ignores the fact that individuals themselves are a product of concrete social conditions. Blacks, and to a lesser extent women and the poor in general, are thus doubly handicapped by social conditions that limit aspirations and achievement and by their own internalized beliefs that their lives are shaped by economic and political institutions beyond their control. It is likely that, if the United States were suddenly to rid itself of racial and sexual discrimination and improve conditions that have traditionally favored white men, some differences between blacks and whites and men and women would persist because disadvantaged groups have internalized beliefs about their lack of personal control.

Researchers should avoid the tendency, however, to see blacks within American society as a monolithic group. Banks (1984) has studied black youths in predominantly white suburbs in the Pacific Northwest and has found a pattern of locus of control that is quite similar to that of the white population. Blacks in this sample were more internal on the test measure than the overall population. They also tended to become more internal as they grew older, and internality was positively related to self-ratings of ability, attitudes toward school, and self-esteem. Although more research is needed, the study suggests that a sizable number of blacks in America are undergoing a process of assimilation in which race is becoming largely irrelevant.

COGNITIVE CONSEQUENCES OF LOCUS OF CONTROL

Numerous studies assessing various aspects of learning have consistently shown that those with an internal locus of control outperform those with an external orientation. In one of the first attempts to validate locus of control, Seeman and Evans (1962) found that hospitalized tubercular patients with an internal perspective knew more about their condition, questioned doctors and nurses more often, and expressed less satisfaction with the feedback they received. In a related study, Seeman (1963) provided reformatory inmates with information on parole, the reformatory, and long-range opportunities, and then tested them six weeks later. Internals recalled significantly more information about parole but only marginally more about the reformatory and long-range opportunities.

Processes underlying the superior performance by internals have also been explored. Internals appear to be more sensitive to potentially useful information and more likely to engage in the initial steps of gathering information. Internals ask more questions (Davis & Phares, 1967) and scrutinize people's faces more often when they behave peculiarly (Lefcourt & Wine, 1969). Internals are not only more alert but also appear to process information more effectively. Internals spend more time making decisions (Rotter & Mulry, 1965), and the amount of time increases as tasks become more difficult (Julian & Katz, 1968). Internals also have more task-relevant thoughts and are less likely to let their attention wander (Lefcourt, Lewis, & Silverman, 1968). Externals tend to be distracted by novel, irrelevant information long after internals have grown accustomed to it.

Internals and externals also differ in their *use of information.* Phares (1968) had subjects learn 10 items of information about four men until they reached effortless recall. A week later he asked them to guess which of eight women and 10 occupations would best suit each. Internals provided 50% more reasons and were three times more likely to give correct reasons to justify their choice. Internals have also been found to be less dogmatic (Sherman, Pelletier, & Ryckman, 1973) and to have fewer irrational beliefs. They are also more persistent and better able to delay gratification (e.g., Mischel, Zeiss, & Zeiss, 1974). In short, internals are more likely to seek out, learn, and retain information. They are more focused, less susceptible to distraction, more persistent, and better able to delay gratification. They spend more time deliberating difficult decisions and perform better than externals even when provided with the same information.

Numerous studies have shown a direct relationship between internality and *academic achievement,* and several reviews of this literature have appeared (e.g., Bar-Tal & Bar-Zohar, 1977; Phares, 1976). Findley and Cooper (1983) have carried out a metaanalysis of 98 relevant studies containing 275 separate comparisons. Of these 275, 70% showed a positive relationship between internality and academic achievement. The average correlation for the general measure was quite low ($r = .18$), but a number of subtrends were uncovered. The relationship between locus of control and academic achievement was stronger for males than for females. The highest correlations occurred for those in junior high, and the smallest were for those starting elementary school or going to university. It is not altogether clear, however, whether age differences reflect real differences in the relationship between locus of control and achievement or are simply due to the different scales used to measure locus of control. The relationship for university students may be weak because they are relatively internal when compared with the general population (Rotter, 1975). The relationship between internality and academic achievement has been found not only in the United States but among Danish children, Hungarian children, and Mexican Americans.

There appear to have been few attempts to directly compare locus of control and *achievement motivation,* but some evidence suggests that they interact. Wolk and DuCette (1973) found that the correlation between achievement motivation and achievement occurred only among internals. More specifically, internals with high achievement motivation were more likely to choose tasks of intermediate difficulty, estimated the probability of success as higher, and performed significantly better on standardized SAT tests and on exams. The authors suggested that some of the inconsistencies in the research on achievement motivation may be due to a failure to consider locus of control. A desire to do well appears to facilitate performance only when there is a perceived connection between outcomes and behavior.

Some qualifications need to be made, however. As already mentioned, the effects of locus of control on cognitive performance are stronger for males than females. Duke and Nowicki (1974) found that there was a direct relationship between internality and grades in boys, but externality was associated with achievement in girls. Prociuk and Breen (1975), using Levenson's scale, found that internal males and females were both more academically successful than those who attributed success to powerful others and chance; but for females, a belief in powerful others was also associated with relatively high grades—that is, the greater the belief in control by powerful others, the higher the level of academic success. These studies suggest that, for females, some belief in an external locus of control may facilitate performance in school. Graham (1989) has recently suggested that there may be differences between blacks and whites as well. These qualifications show the danger of treating locus of control as a unitary dimension in which internality is good and externality is bad. Rotter (1966) initially speculated that there was a curvilinear relationship between locus of control and psychological adjustment. Individuals at either extreme were seen as less adapted. Internals do possess negative qualities: They are less tolerant of both themselves and others; they take more responsibility for both success and failure and are, therefore, more self-critical when they fail; and they also perform less well when exposed to uncontrollable outcomes or extended failure.

But, for the most part, internals excel. Not only do they seek out and process information more effectively, but they perform better in school. Their superior performance extends even into graduate school where they are more likely to complete the requirements and obtain their PhD (Otten, 1977). Internals react less strongly to negative events (Kobasa, 1979) and are more likely to use preventive and corrective measures to facilitate health and recovery (Strickland, 1978). Externality has been linked to suicide-proneness, depression, and learned helplessness. Even sexually, internals outperform externals and have more frequent and more satisfying sexual encounters.

Locus of control has been found to affect a variety of different cognitive activities, but it is possible to overestimate its contribution. Locus of control is only one of many factors that help determine performance. Naditch and de Maio (1975) obtained self-reports of interest and competence in four areas—school achievement, social life, sports, and activities at home. Among boys, locus of control was related to academic achievement but only for those in the top third of their class. Similar trends occurred for social competence and sports. In each case, locus of control was associated with high levels of ability but only when the activity was highly prized.

It is rather astounding that an acquired belief, such as locus of control, can have so many cognitive consequences. It helps determine how we seek out, learn, and use information and affects our ability to perform in specific areas. Although locus of control is only one factor influencing performance, it has been found to have a low-to-moderate effect on a wide range of cognitive tasks. Like achievement motivation, it appears to be almost entirely learned, yet it affects virtually every aspect of our lives. As we discover differences between groups—male-female, black-white, rich-poor, and so on—it is hard to escape the impression that entire segments of our society are being saddled with an unmistakable handicap that limits performance on a wide range of tasks.

Some evidence, however, indicates that locus of control can *change*. As noted previously, substantial changes occurred during the 1970s for the American population as a whole, and there appears to be a shift back toward the internal perspective during the 1980s (Strickland, 1989). Increases in an external locus of control have also been associated with early childhood trauma and even such major events as the government-initiated lottery that determined eligibility for the draft (McAuthur, 1970) or the defeat of the liberal candidate, Eugene McCarthy, at the 1968 Democratic Convention (Gorman, 1968). Doherty (1983) found that divorce has a temporary effect on externality in women. Women who are physically abused by their husbands also become more external than nonbattered wives, and the degree of externality increases as long as they remain in the abusive relationship (Cheney & Bleker, 1982). Rotter (1975), himself, suggested that the surprising popularity of the locus of control concept may be related to persistent social problems that confronted Western countries during the 1970s and the subsequent feelings of powerlessness and alienation.

Shifts toward the internal end have also been reported. Harvey (1971) found that the longer a person held an administrative post in the upper echelons of government, the more internal they became. Andrisani and Nestel (1976) carried out a two-year longitudinal study of the relationship between locus of control and success in the labor force. They found that reentry into the labor force, promotion, and increased salary were associated with an increase in internality, whereas negative experiences produced a more external perspective. Remanis (1971) gave the Battle–Rotter cartoon test to children and assigned the most external to experimental and control conditions. He then held weekly meetings with teachers of those in the experimental group in which he discussed procedures for encouraging internality. Those in the experimental group showed a significant shift in locus of control and demonstrated more behaviors associated with an internal perspective such as increased interest and classroom participation.

Spence and Spence (1980), on the other hand, found an immediate shift in locus of control and self-esteem among adolescent male offenders following social skills training, but the effect was short-lived. Young offenders in an institutional setting quickly returned to their pretreatment levels on both measures. It seems that people's level of locus of control, like their level of achievement motivation, is not only the result of social conditions but is maintained by social conditions, and it is unlikely to undergo a radical transformation when people do not have significant control over their own lives. This can be seen more clearly in the related concept of learned helplessness, which has been studied extensively in recent years.

LEARNED HELPLESSNESS

Although most people think of learning as the acquisition of knowledge or skills that improve performance, it is also possible to learn maladaptive behavior or even to learn not to respond at all. An extreme form of externality occurs in the case of *learned helplessness,* which is a devastating sense of not being able to control what happens in one's life (Seligman, 1975). A large body of empirical research suggests that both people and other animals can learn that their behavior has no effect on the environment and that aversive events cannot be avoided or escaped. Faced with inescapable stress or pain, they no longer resist and simply give up. People who have repeated experiences with events that are outside their control develop a general expectancy that future events will be uncontrollable as well.

At the heart of the original theory was the view that learning that outcomes are uncontrollable leads to three deficits. First, *motivation* is lost. People who are helpless typically do not try to learn new behaviors because they believe they no longer control what happens to them. Lack of motivation produces *cognitive* defects. People fail to learn from experience and overestimate the amount of failure and their lack of control. Finally, learned helplessness leads to an *emotional* deficit. The helpless person becomes sad and apathetic and typically undergoes a loss of self-esteem. Other symptoms include self-blame, feelings of inadequacy, loss of appetite, insomnia, and a loss of interest in what is going on. Seligman (1975) feels that learned helplessness is at the core of many symptoms typically associated with *depression.*

Since learned helplessness occurs in both humans and animals, it is not necessary to postulate an extensive set of beliefs to account for it, but more recent evidence suggests that, in humans, attributions for success and failure can make a difference (Abramson, Seligman, & Teasdale, 1978). When people learn that desired outcomes and behavior are independent or that negative outcomes cannot be avoided, they make *attributions* about the cause to estimate the probability of success in the future. If future success seems highly unlikely, then they develop symptoms of helplessness and become depressed. Attributions for behavioral outcomes are based on three separate dimensions—internal-external, stable-unstable, and global-specific. The internal-external dimension refers to whether the individual perceives the cause of behavior as residing within the self or within the environment. Stable-unstable is based on the perceived duration of the event—long-lived and recurrent versus brief and intermittent. The global-specific distinction refers to the belief that an event influences just one aspect of behavior or a broad range of situations.

In the new theory, depression is most severe when negative events are attributed to internal, stable, and global factors. Failure at a task, for example, has a more negative effect when it is attributed to stable, internal factors, such as lack of ability; and it is even more devastating when it is seen as a general lack of ability. Those who attribute unpleasant outcomes to personal inadequacies blame themselves and experience guilt, depression, and a loss of self-esteem. Individuals have a characteristic *cognitive style* for explaining bad events, and those with an internal, stable, and global style are more prone to depression. They respond to negative events by blaming themselves and assume that their personal shortcomings will last forever and affect everything they do. Seligman and his associates do not claim that these attributions always occur consciously. Under some circumstances, people may quite self-consciously query themselves about why something occurred and use their conclusions to judge the probability of future success. In other cases, implicit inferences drawn from similar cases in the past influence the expectation of doing well. If attributions were based on conscious processes, then the method of introspection would be the best way to study them, but since they are not, a variety of indirect methods must be used.

A great deal of research supports the claim that a negative attributional style is linked to depression (see Brewin, 1985; Coyne & Gotlib, 1983; Peterson & Seligman, 1984, for reviews). Sweeney, Anderson, and Bailey (1986) carried out a metaanalysis of 104 studies involving nearly 15,000 subjects which showed that, for negative events, attributions to internal, stable, and global causes were reliably and significantly associated with depression. For positive events, external, unstable, and specific attributions were linked to depression but the degree of association was much weaker.

One of the least explored areas is the *origins* of the depressive explanatory style. Peterson and Seligman (1984) have suggested three possible sources. A child may learn helplessness by imitating the parents, particularly the mother. They found significant correlations between mothers and children for both explanatory style and depression. Second, criticism from teachers may also play a role (see Dweck & Licht, 1980). Finally, both explanatory style and the level of depression depend, at least in part, on traumatic experiences.

More recently, Seligman (cited in Buie, 1988) has argued that recent increases in the level of depression can be traced to a lifestyle that stresses individualism rather than commitment to others. He noted that several studies sponsored by the Alcohol, Drug Abuse and Mental Health Administration found that people born after 1945 are 10 times more likely to suffer depression than those born 50 years earlier. Seligman dismisses the idea that this increase is due to a genetic-environmental interaction because the rate of increase is simply too high. He attributes the increase to the decline in patriotism following the political assassinations, Watergate, and the Vietnam war; and to a breakdown in the family and religion. Without such institutional buffers and community support, small setbacks become much more devastating. He concludes that a necessary condition for a meaningful existence is an attachment to something larger than oneself.

The close link between learned helplessness and externality is sometimes obscured because learned helplessness is based on internal attributions. Internal attributions occur during unpleasant experiences and focus on personal inadequacies, such as lack of ability. During positive outcomes, internal attributions actually help relieve the sense of helplessness. If a person cannot avoid aversive events, then the only possible control resides *outside* the individual. Although the overlap is far from perfect, learned helplessness can be seen as an extreme form of externality based on the belief that the individual is personally powerless to control what happens. The belief that behaviors and outcomes are independent decreases the likelihood of initiating and sustaining activity and therefore decreases the likelihood of success. The link between these two concepts suggests that they may share many of the same antecedents and produce many of the same effects.

Learned helplessness, like locus of control, can be seen as a relatively stable personality trait that can be changed under some conditions. Although the social conditions responsible for learned helplessness have not been well explored, Seligman (1975) suggests that there is a link between learned helplessness and poverty and that extreme poverty can produce a sense of helplessness from which few escape. No treatment of learned helplessness would be complete, however, without some discussion of those who do escape. These have been described as *resilient children,* and they appear to have a special kind of buoyancy and an ability to bounce back from even the most adverse circumstances.[4] Some even have the capacity to harden under stress, which has been described as a "steeling effect" (Murphy, 1974; Rutter, 1985) and show a "precocious maturity" in their relations with other people. As Garmezy (1971) has put it, "With our nation torn by strife between races and social classes, these 'invulnerable' children remain 'keepers of the dream'" (p. 114).

It would be wrong, however, to equate learned helplessness or an external locus of control with some kind of character flaw and assume that resilient children somehow have the strength of character to overcome social circumstances. Resiliency itself is a result of social conditions and almost invariably includes strong support from some significant other. Werner (1986) stresses the importance of early childhood experiences, such as plenty of attention and no prolonged separation from the caretaker during the first two years. Lack of parental conflict and the absence of siblings during this period also seem important. When stress does occur, it tends to be limited and is usually followed by restoration of previous conditions or a change in circumstances for the better (Rutter, 1978). Resilient children also receive a great deal of attention from someone who attempts to make them feel special, bolster their self-esteem, and provide them with a sense of social values and self-worth. As a result, a (yet to be determined) proportion of children develop the capacity to rise above their adverse conditions and may well be the "keepers of the dream."

Wilson (1987) has pointed out that the poor in the United States are not a monolithic group. The considerable variations among the poor suggest that they would respond differently to social change. Some have had only periodic experience with poverty, whereas others have been persistently poor for several generations. Some are upwardly mobile; others are on their way down. Some embrace middle-class values; still others identify with the working and nonworking poor. Households also vary in the extent that they are embedded within an extended family or are part of a network of mutual support. These and other factors help determine how much each individual believes or can be led to believe that personal involvement can affect their life or change the world.

[4] This treatment of resilient children draws from an unpublished paper by M. Bookman (1990) that examines resiliency among children with alcoholic parents. The focus of the current chapter and my interpretation of the data are somewhat different.

CHAPTER 5

Level of Aspiration

After a general introduction, the relationship between self-esteem and level of aspiration is explored in detail. This relationship is used to explain the failures of desegregation in the United States. The chapter ends with a brief discussion of the recent concept of self-handicapping.

We have seen how achievement motivation and locus of control help determine the level of task difficulty selected by people: Those with high achievement motivation prefer tasks of intermediate difficulty, whereas those with low achievement motivation or high fear of failure select tasks that are either very difficult or extremely easy. But people's level of achievement motivation is also determined more directly by their reference group and the people with whom they closely identify. The level of aspiration can improve performance if the group sets high but realistic standards, but it can also lower performance when the standards are extremely low. In the latter case, people are caught in a conflict between a desire to achieve and the threat of social isolation. Much of the research on the role of social models on the level of aspiration can be traced back to the work of Kurt Lewin (1890–1947) and his students at the University of Berlin.

Lewin's (1935) *field theory* stresses goal-directed behavior and much of his early research focused on what happens when such behavior is interrupted. The concept of *level of aspiration* was introduced by Tamara Dembo (1931) in her study of frustration and anger. She speculated that anger occurs when people cannot achieve their goals and found that the amount of anger was based on the intensity of the drive rather than its importance. People may become extremely angry when a trivial goal is frustrated if they are sufficiently involved. After Lewin came to the United States in the early 1930s, his interests began to shift from intrapsychic processes occurring "within" individuals to social psychology and group dynamics. This shift in interest was based on a number of factors, but one of the more important was the growing recognition of the role of others in setting goals and creating conflict, frustration, and aggression. Other people serve as models, as objects of interest, as obstacles, and as targets for aggression.

Lewin's interest in group dynamics was also based on a growing concern with social problems, which were particularly pronounced during the 1930s. He felt that many problems, such as poverty, racism, anti-Semitism, and aggression, were problems that occurred within and between groups and that, at times, it was easier to change group behavior than that of individuals. Groups provide a means for social comparison, a forum for discussion and a source of social support so that attitudes developed within groups are more resistant to change. As an outsider, Lewin was able to look at the United States with a detached perspective. He was greatly impressed by its egalitarian and democratic values but also deeply disturbed by its racism and negative attitude toward Jews. Lewin's interest in group dynamics helped inspire a number of social psychological theories during the postwar period, but the one most closely associated with level of aspiration is Leon Festinger's (1954) social comparison theory.

SOCIAL COMPARISON PROCESSES

The roots of Festinger's social comparison theory can be seen in his earlier research on informal communication (Festinger, 1950, 1951). In these papers, Festinger argues that group communication is based partly on *pressures toward uniformity.* The amount of communication is determined by the amount of discrepancy, the relevance of the issues, and the level of group cohesion. Communication is directed, at least initially, toward those with different opinions, and it continues as long as there is a chance that they will change their beliefs. People's ability to resist these pressures depends on their desire to remain in the group and also varies with the extent in which their beliefs are anchored in other groups or serve important personal needs. But groups have a tendency to maintain pressure, and if change does not occur, those with different opinions are often rejected. Expulsion of people with different opinions helps maintain uniformity, and the threat of expulsion is an important source of motivation that can be used to bring people into line.

By 1954, Festinger's research on group dynamics, communication, and cohesion provided the basis for a new theory of social comparison that attempted to bring together and integrate a considerable body of experimental research. His complete statement is somewhat detailed and consists of nine formal hypotheses plus their corollaries and derivatives. The basic assumption is that people have a need to evaluate their beliefs and level of ability (Hypothesis 1), and when objective evidence is not available, they tend to use other people as a source of social comparison (Hypothesis 2). Two corollaries of the second hypothesis are that, when objective evidence is available, social comparison is unnecessary, and when neither physical nor social evidence is available, people tend to be unsure about the accuracy of their beliefs. People prefer objective evidence whenever they can obtain it. Some abilities, such as weight lifting or running, can be easily measured but others, such as intelligence, are more abstract and depend on the judgments of others. The search for objective measures helps explain the popular appeal of psychological tests or even pseudo-psychological tests in popular magazines that purport to measure everything from marital satisfaction to sexual appeal.

Festinger's third hypothesis is based on the *similarity* of those who serve as a source of social comparison. In general, people tend to prefer others who are similar. We compare ourselves with others who hold similar beliefs because consensus within a group tends to confirm the correctness of our views. We also compare ourselves with others who have roughly the same level of ability because this provides more information about our relative performance. Weekend skiers, for example, compare themselves with other weekend skiers rather than novice or Olympic contenders. The pressure toward uniformity tends to draw like-minded people and those with similar levels of ability, and communication within groups tends to reduce existing differences and makes group members more and more alike.

Festinger's fourth hypothesis focuses on *differences between opinions and abilities.* There is a unidirectional upward drive for abilities (where people strive to do better and better) that is largely absent in the case of opinions. For this reason, attempts to reduce differences in opinions are based simply on pressures toward uniformity, but for abilities, pressures toward uniformity compete with the drive to improve. The net result is that group members strive to improve just up to a point where they are slightly better than other members of the group. The group serves as a powerful anchor that limits the level of aspiration, particularly when the group is cut off from other groups.

Nonsocial factors also make it difficult or impossible to improve beyond a certain point (Hypothesis 5). Someone who wants to be a really good downhill skier may lack the balance and coordination necessary to compete with the very best. As a result, a person will tend to associate with others who have similar levels of ability and will restrict his or her performance by eliminating those who perform much better. A person whose level of ability significantly exceeds that of the group can reduce perceived differences either by limiting their own performance or devoting time and effort toward improving the performance of others. Festinger's other hypotheses suggest that when people stop comparing their performance with others, they tend to derogate those others and feel hostile (Hypothesis 6) and that any factor increasing the importance of the group, such as group attractiveness or topic relevance, will increase pressures toward uniformity (Hypothesis 7).

Although Festinger's social comparison theory was developed to explain processes occurring within groups, it applies to differences *between* groups as well. People tend to move into groups with similar beliefs and similar levels of ability. The movement into and out of groups allows society to maintain a wide range of opinions and accommodate people with different levels of ability. Differences in level of ability give rise to status differences within society, and these differences are maintained not only by those with the highest level of ability but by those with lower levels because it allows those with less ability to ignore groups whose level of ability is significantly greater than their own.

For our purposes, the most significant aspect of social comparison theory is the way it can be used to explain individual differences in levels of aspiration. People tend to use others who are similar or have similar levels of ability as a source of social comparison. The threat of ostracism tends to restrict the performance of those with high ability, and performance levels are typically anchored somewhere around the mean. This anchoring can help explain some of the differences in academic performance between blacks and whites and, ultimately, help throw some light on the failures of desegregation in America.

ROLE MODELS IN AN UNDERCLASS SOCIETY

The changing demographics among blacks in the United States show that one of the most disturbing trends is the movement of middle-class blacks out of the ghettos and the creation of an underclass society. As mentioned previously, inner cities previously contained a "vertical integration," in which lower- and middle-class families lived together, shopped together, and sent their children to the same schools. The presence of well-educated professionals served as a social buffer that helped provide stability and absorb the shock of poverty and unemployment. The movement of middle-class blacks out of the ghettos makes it difficult to maintain basic institutions, such as churches, recreational facilities, and schools, and as these deteriorate, there is a corresponding drop in community spirit, neighborhood identification, and social sanctions that discourage deviant behavior and crime.

Such changes also alter the role models available to lower-class blacks. The absence of well-educated professional people within the community means that lower-class children are likely to have few examples of people who have achieved success through education. As mentioned in Chapter 4, poor blacks are far more likely than poor whites to live in extreme poverty areas (where the rate of poverty is above 40%) (Wilson, 1987). Poor blacks are also far more likely to be cut off from other groups and experience what has recently been described as "hypersegregation" (Massey & Denton, 1989).

Massey and Denton (1989) used five dimensions—evenness, exposure, clustering, centralization, and concentration—to compare black and Hispanic segregation in 60 American cities. *Evenness* refers to the extent that minority members are spread evenly across the city or, more formally, the proportion of people who would have to be moved to achieve an even distribution. Highly segregated cities tend to have an uneven distribution in which minorities are restricted to a few local areas. *Exposure* is the degree of potential contact between groups. It is calculated by an isolation index that measures the probability that members of different groups will share the same area. *Clustering* is the extent that segregated areas join each other and form one large, continuous ghetto. Such clustering leads to a higher degree of social isolation because minority groups are further separated from members of the majority. *Centralization* refers to the degree that minority members are settled around the center of the city,[1] whereas *concentration* is the relative amount of physical space occupied by each member of a minority group or the total share of the urban environment.

These researchers found that blacks were not only more segregated on every single dimension but were also far more likely to be segregated on all five dimensions at the same time. Six cities (Baltimore, Chicago, Cleveland, Detroit, Milwaukee, and Philadelphia) were highly segregated on all five dimensions, and four others (Gary, Los Angeles, Newark, and St. Louis) were highly segregated on four of the five dimensions. These 10 cities contain 29% of all metropolitan blacks and 23% of all blacks living in the United States, which means that almost a quarter of all American blacks live in densely settled urban centers cut off from other communities and clustered around a central core. People in such communities rarely get a chance to meet or interact with members of other races. When more liberal criteria were used to determine "low" levels of segregation, nine smaller cities qualified but these contained less than 2% of the total black population.

When Hispanics were compared, in contrast, none of the dimensions were high or even moderately high by black standards. Only four cities (Chicago, New York, Newark, and Patterson, New Jersey) were high on as many as three dimensions at the same time. Three of these cities contain large numbers of

[1] Centralization is a concept developed at the University of Chicago to explain conditions in cities like Chicago. It describes cities in terms of concentric rings and assumes that there is a downtown business and financial district surrounded by a derelict region that has yet to be developed. Outside this is a band of working-class neighborhoods and further out are middle-class suburbs. The rate of crime and social disruption increases as one moves in toward the center of the city and decrease when moving outward and this trend seems to be completely independent of whatever ethnic group surrounds the central core. Many cities, such as Cleveland and Chicago, fit this pattern. Others with sizable black populations, such as Boston, New York, and Washington, do not. These cities (and others like them), nevertheless, have highly segregated and well-defined black communities—Roxbury, Harlem, and the East End, respectively. The use of centralization as a criterion suggests that Massey and Denton may have *underestimated* the extent of hypersegregation and that many more blacks may be cut off from other groups.

Puerto Ricans, who traditionally receive more discrimination than other Hispanics (Massey & Bitterman, 1985). Moreover, those cities with the highest concentration of Hispanics—Los Angeles, Miami, San Antonio, and San Diego—were not highly segregated on *any* dimension.

In short, hypersegregation appears to be limited to blacks. No other group experiences the same degree of social isolation. Blacks occupy a unique and distinctly disadvantaged position within the United States because they are not only cut off from resources such as education and employment but they are also increasingly cut off from those who have access to such resources and might otherwise serve as role models.

The problem, however, is not limited to ghetto communities. Ogbu (1986) points out, "Children respond positively to school IF they observe that older people in their community usually obtain jobs, wages and other social benefits commensurate with their level of schooling" (p. 40). Unfortunately, such is not the case with many blacks in America. The percentage of blacks in colleges and universities more than doubled during the 1970s but this had little if any effect on the overall level of black unemployment. In 1977, the majority of unemployed black youth had either completed high school or gone on to college (Pinkney, 1984).

With any level of education, blacks have more difficulty obtaining work than whites, and until recently, the gap *increased* with the level of education. Recent census data, for example, show that while the overall ratio of unemployed black to white high school dropouts was two to one, the ratio among those who had finished school was three to one. Blacks are also far more likely to be underpaid or overqualified for the jobs they do have. Such treatment may lead to what Ogbu (1986) calls a "collective institutional discrimination perspective," which is a belief that it is difficult to advance within mainstream society by obtaining an education or behaving like whites. Although blacks appear to value education as much as whites, they are less likely to see education as a means for obtaining high-status jobs.

It is often assumed that educational aspirations are mediated by *self-esteem* and that lower-class blacks, because they experience poverty and discrimination, have low levels of self-esteem. Many people simply assume that groups victimized by prejudice and discrimination have low self-esteem. Hundreds of studies designed to explore the relationship between minority status and self-esteem, however, have found little or no relationship. Many of the better controlled studies have actually found that minority group members have higher levels of reported self-esteem than members of the majority. The same is true for *socioeconomic status,* which overlaps race to a great extent. Although some studies have found a direct relationship between socioeconomic status and self-esteem, others have found either no relationship or an inverse relationship, where poorer children have higher levels of self-esteem.

There are a number of reasons this relationship fails to occur. One is because people's level of self-esteem develops very early and is relatively resistant to change. Parents have an enormous impact on their children's self-esteem because they are their first social contact, serve as powerful role models, and interact frequently. In general, warmth and affection is positively related to self-esteem, whereas neglect and abuse lead to low levels of self-esteem. Some of these effects can be extremely subtle. Murray and Trevarthen (1985), for example, have studied interactions between mothers and their 2-month-old infants. Otherwise attentive mothers were instructed to freeze their behavior and become immobile for two minutes. The infants became distressed, avoided eye contact and gradually withdrew from the interaction. These same responses do not occur at 5 months of age, and there seems to be something like a critical period in which warm relationships lead to what Erikson (1963) has called "trust versus mistrust." Researchers can only speculate about the repeated impact of extensive parental neglect on growing children's sense of self-worth.

Although it seems plausible that self-esteem is based on knowledge about the self, there is considerable evidence that self-esteem develops earlier and then helps shape a person's self-concept. Even basic distinctions based on race or gender are not made until the second or third year, and they are not recognized as permanent until much later (Kohlberg, 1966). Young children often believe they can change their sex by cutting their hair or changing the way they dress, but by the age of 3 or 4 years, they have had extensive social contact and their self-esteem is well established. Although this may change later on, it helps determine how or even whether information about the self is processed.

A second reason is that the overall level of self-esteem is based on *social comparison.* Members of minority groups compare themselves with the people they most frequently associate with (Rosenberg, 1981). The self-esteem of black children is strongly related to what they believe their friends and parents think of them. Since they are frequently segregated, their self-image is based on the reflected appraisal of other members of their own group. Ghetto children have less contact with wealthier people and are therefore more insulated in their social comparisons. Because all their friends are poor, poverty is more or less taken for granted.

A third, somewhat related, reason is that young children have a *poor understanding of socioeconomic status.* Rosenberg and Pearlin (1978) studied children and adolescents and found that less than a quarter of those between 8 and 11 knew children who were richer than themselves. They also found that young children were less likely to have heard of the term "social class." If they had heard it, they were unlikely to have formed an opinion of where they belonged, and when they had an opinion, they were typically wrong. When compared with wealthier white children, the level of self-esteem also depends on who receives the blame. Blacks who blame the system are far less likely to experience low self-esteem than those who blame themselves.

People also tend to select the characteristics and traits on which they base their self-esteem. The overall level of self-esteem is based on the general evaluation people have of themselves—whether or not they see themselves as competent, likable, or lacking in important personal attributes—but whether these evaluations are high or low depends on the relative importance of individual traits. Rosenberg (1965) points out that people tend to value those things at which they excel and devalue those things at which they do poorly. This leads to the interesting paradox that almost everyone can feel superior to others in some way.

Once self-esteem has been established, it is relatively difficult to change. People may go through periodic mood swings in which they feel good or bad about themselves but the overall level of self-esteem remains relatively stable. Most people have moderately high levels of self-esteem, and they maintain it in a number of different ways. They associate with people who like and respect them and avoid those who have a poor opinion of them. They place a high value on tasks and activities at which they excel and devalue those things at which they do poorly. People can also distort or reinterpret the negative information they do receive. This tendency to distort information to maintain high levels of self-esteem can be seen in a survey of more than 2,000 American boys in the 10th grade (Backman, 1970). When asked to rate themselves in terms of academic ability, 83% rated themselves above average.

Finally, there may be an *inverse relationship* between level of aspiration and self-esteem. Self-esteem can be measured in a number of different ways. One approach uses direct questions in which people agree or disagree with statements, such as "I certainly feel useless at times," "On the whole I am satisfied with myself," or "I have a number of good qualities" (Rosenberg, 1965). The overall score is simply the sum of individual items suggesting high self-esteem. A second approach is based on the discrepancy between the perceived self and the ideal self (Cohen, 1959; Sherwood, 1965) and it draws from William James's (1890) famous formula:

$$\text{self-esteem} = \text{success/pretensions}$$

in which the level of self-esteem is based on the current level of achievement and what the person would like to achieve. In this approach, people provide two descriptions, one for their actual self and one for the person they would like to be. The overall level of self-esteem is based on the discrepancy between these two descriptions—the greater the discrepancy, the lower the self-esteem.

James (1890) pointed out that this formula leads to "the paradox of a man shamed to death because he is only the second pugilist or the second oarsman in the world. That he is able to beat the whole population of the globe minus one is nothing: he has 'pitted' himself to beat the one, and as long as he doesn't do that nothing else matters" (p. 310). This formula helps to explain why people can maintain high levels of self-esteem with low levels of achievement. If expectations are low then they can be easily realized. Indeed, one of the most common defensive strategies appears to be based on underachievement. People deliberately strive for goals which can be easily achieved in order to avoid the negative consequences associated with failure.

Not everyone agrees that race and self-esteem are unrelated. Most of the research reporting high levels of self-esteem among blacks has tended to use self-reports, and clinicians working directly with black children have expressed skepticism about whether verbal statements of high self-esteem reflect their true feelings. Spurlock (1973), for example, claims that statements suggesting "black is beautiful," which is most common among lower-class blacks, is a reaction formation used to hide a negative self-image. Mamie Clark (cited in Pouissant, 1974) has also suggested that there has been very little real change in self-esteem among blacks since she began her research in the 1930s.

Epstein (1985) argues that the need for self-esteem stems from an internalization of the child's need to be loved by his or her parents. Children learn to view their behavior from their parents' perspective and develop a similar attitude toward it. He points out that self-esteem, at this level, is not based on a conscious self-evaluation. It is one thing for a person to describe him- or herself as high in self-esteem and another to possess the "quiet confidence, emotional stability, frustration tolerance, openness to new experience, and lack of defensiveness that can theoretically be expected of people with high self-esteem" (p. 302).

It is important to note that hypersegregation is widespread but by no means universal. As mentioned in Chapter 4, one of the most significant trends in recent history is movement of middle-class blacks out of the ghetto into formerly white middle-class neighborhoods where conditions are quite different. There are also important *gender differences* within the black community. Black parents tend to have higher educational aspirations for their daughters. This has led to what Bock (1969) has described as "the farmer's daughter effect," since many farmers also encouraged the education of their daughters rather than their sons in order to keep the sons on the farm. Smith (1982) reviewed the literature and concluded that "black female adolescents generally have not only higher aspirations but also greater expectations of completing their goals" (p. 272). Black females also tend to do better in school and are less likely to drop out. Data from the Department of Education for 1984 showed that there were 200,000 more black women than black men in institutes of higher learning and 160,000 more black female undergraduates (Irvine, 1990).

Dweck and Bush (1976) have suggested that boys in general experience more pressure to conform and that this *peer pressure* tends to work against educational aspirations. Boys are far more likely to experience a conflict between peer values and adult values and are often forced to violate adult norms in order to be accepted. There is often a counterculture among boys based on rule breaking and group solidarity, which limits educational aspirations and distinguishes "typical" boys from high achievers. The terms used to describe children with a genuine interest in school, such as "suck ups," "brown noses" and so on, suggest that there is something effeminate, possible homosexual, about doing well. These differences are not limited to academic programs. They occur in vocational and trade schools as well (Sennett & Cobb, 1972).

Girls, on the other hand, often confront a single standard that does not conflict with that of adults. Female students can maintain their popularity while trying hard and doing well in school. These differences are by no means limited to blacks. A similar conflict seems to exist among lower and middle-class white boys and in other cultures as well (Bronfenbrenner, 1970). It seems to be a widely practiced self-imposed limitation based on a need for social acceptance that limits educational aspirations among males.

According to Irvine (1990), gender differences in levels of aspiration also occur because of more general attitudes within the black community. Black girls are often given a great deal of early responsibility in the home. They are expected to help out with the housework, take care of younger children, and so on. They are also taught both verbally and by example not to depend on males. Boys, on the other hand, often receive a mixed message. On the one hand, traditional sex attitudes stressing masculinity and femininity are very strong among blacks. Males are expected to show strength and courage, to be assertive and stand up for themselves. On the other hand, many poor black parents know that these traits will likely get their sons into trouble and possibly killed. The end result is that black females are encouraged to be strong and independent and do well in school, whereas black males are discouraged from being overly assertive.

As a result, black females tend to do better in school, *at least initially.* By the time black males and females enter university, males typically have higher levels of aspiration. Again, the pattern can be traced to traditional sex attitudes within black communities. Whereas black females are expected to be, and indeed must be, strong and independent, positions of leadership and high-status jobs are typically reserved for men. The civil rights movement, for example, was dominated by black men. Black males are also far more likely to become preachers, lawyers, and doctors. These differences occur in spite of the fact that black women are perceived as less of a threat and are frequently encouraged by the white community. Because there are fewer black men than women in colleges and universities, those who do make it tend to be a more select group.

Prejudice, discrimination, and lack of job opportunities tend to produce hostility and distrust toward the school system making it hard for black children to accept the goals of education. As a result, blacks often adopt various *coping strategies* designed to succeed in nonacademic areas and maintain self-respect and self-esteem (Ogbu, 1985, 1986). These include collective struggle, clientship, and hustling. Collective struggle involves a deep commitment to social change and a belief that problems can be reduced through collective action. In the strategy of clientship or "Uncle Tomming," black people attach themselves to white patrons to obtain jobs, promotions, and other benefits. Deferential and subservient behavior typically accompanies such attachment. Hustling occurs when people take advantage of the underground street economy to make money without working for others. Some of these activities, such as selling drugs and prostitution, are illegal, but others, such as playing pool or gambling, are not. Hustlers are often seen as "street smart" and they are frequently admired for their high levels of intelligence, but they seldom have high levels of education. They serve as alternative role models who have succeeded without becoming a part of mainstream society. Many inner-city black children who do not find school meaningful or interesting become part of the street culture when they are between 8 and 12 years of age (Ogbu, 1985).

The most extreme response is what Ogbu (1985, 1986) calls "cultural inversion," which is a complete avoidance of anything associated with white society. Many inner-city blacks come to define black culture in opposition to white culture and develop unique styles of walking, talking, and dressing. Some activities, such as sports, are considered legitimate forms of expression, but school itself is often regarded as a white activity and blacks who excel academically are accused of acting white. Although there are important exceptions, prejudice, discrimination, lack of successful role models, and cultural inversion are sufficiently common to discourage the development of high standards in a significant number of black children. As a result, black students often fall further and further behind white students of the same age. These problems do not disappear when black children attend integrated schools. In fact, the gap between blacks and whites has increased since schools were desegregated. There is also a fairly high degree of self-segregation in integrated schools that prevents blacks and whites from interacting or using each other as a source of social comparison.

THE FAILURES OF DESEGREGATION IN THE UNITED STATES

When the Supreme Court of the United States unanimously decided to eliminate government-enforced separate educational facilities in the 1954 case *Brown v. the Board of Education of Topeka, Kansas,* it based its position partly on a statement prepared and signed by 32 anthropologists, psychiatrists, psychologists, and sociologists (1953). The statement began with a general discussion of the effects of segregation, prejudice, and discrimination and asserted that, during their development, black children learn from the environment that they belong to a group with inferior status in the United States. They react to this knowledge with humiliation and a sense of inferiority that leads to self-hatred and low self-esteem. The effects are most strong for those in the lower socioeconomic classes. Middle-class and upper-class black children are more likely to withdraw, become submissive, or show an exaggerated conformity to middle-class white norms. Another consequence of this knowledge was said to be a drop in the level of educational aspiration leading to lower levels of performance and subsequently to poverty, unemployment, and crime.

Several studies were cited showing that very young black children were already aware of the status differences between whites and blacks. The most famous of these, carried out by Kenneth and Mamie Clark (1939, 1947), examined the racial attitudes and self-perception of black children between 3 and 7 years of age. They found that black children who were given a choice between a black doll and a white doll usually preferred the latter. When they were asked which was prettier or nicer, they consistently selected the white one. And when they were asked which one was like them, many (particularly those in the North) burst into tears (K. Clark, 1980). The Clarks' research seemed to document the self-hatred and rejection experienced by blacks in the United States and along with other studies, it was used by the Supreme Court to support the claim that separate educational facilities are inherently unequal and are therefore a violation of the right to equal protection of the law guaranteed by the Fourteenth Amendment.

Because one of the primary purposes of desegregation was to raise the levels of aspiration and self-esteem among blacks, a large number of studies have subsequently been carried out to see if changes have actually occurred. Unfortunately, much of the research on self-esteem is inconsistent and difficult to evaluate. Some have shown increases; others have shown decreases; and still others have shown no particular pattern. Later replications of the Clarks' study conducted during the 1960s and early 1970s revealed a reversal of the original findings (Brand, Ruiz, & Padilla, 1974), but whether this actually reflects an increase in self-esteem is difficult to tell.

As mentioned previously, it is not at all clear that blacks in segregated communities experience low self-esteem or that there is a direct relationship between self-esteem and level of aspiration. Rosenberg (1979) has found evidence that blacks in all-black high schools have higher levels of self-esteem, on the average, than blacks in predominantly white high schools. He attributes these differences to social comparison processes. Blacks in segregated schools compare themselves with other blacks, whereas those in integrated schools use both black and white students as a source of social comparison. Other factors include teasing and mistreatment by white students and a greater awareness of socioeconomic differences and family disruption (Rosenberg & Simmons, 1972). Similar differences have been found for Jews, Protestants, and Catholics in segregated and integrated communities.

As for aspiration levels, all of those who have reviewed the literature have found that levels of aspiration among segregated blacks were at least as high as those among segregated whites (e.g., Epps, 1975; Proshanky & Newton,

1968; Weinberg, 1977). St. John (1975) examined changes reported in 25 studies of desegregated schools and found that the level of aspiration either decreased or remained the same. Again, the amount of decrease was directly related to the proportion of white students in the classes. As the proportion of white students increased, the level of aspiration among blacks decreased.

The effects of desegregation on self-esteem and level of aspiration should be kept separate, however. It is possible that blacks in segregated schools have high levels of self-esteem and low levels of aspiration. Indeed, blacks seem to rely less on academic performance as a source of self-esteem. Rosenberg and Simmons (1972), for example, found that failing grades were highly related to low self-esteem among white students, but the association was much weaker among blacks. Academic competence is only one aspect of global self-esteem, and it appears to play a smaller role among black students.

Also as mentioned previously, there is evidence that self-esteem and level of aspiration are sometimes *inversely related* (Cohen, 1959; Sherwood, 1965). Self-esteem is a function of what a person has accomplished and what he or she wants to accomplish. A person may have high self-esteem either because of having accomplished a great deal or because of expecting very little. Self-esteem is easy to maintain when aspirations are low. Carl Rogers (1967) has found that differences between the perceived self and the ideal self typically diminish during the course of psychotherapy, and this change is accompanied by increases in self-acceptance and self-esteem.

Black children often express high levels of aspiration, saying that they would like to be doctors, lawyers, or politicians, but these are not systematically related to academic achievement. The goal of being a doctor, for example, does not suggest that a person should study mathematics or science particularly hard to get into medical school. These professions are chosen because they are highly visible within the black community and bring high wages and status. According to Jacqueline Irvine (personal communication, February 1990), black children are frequently asked what they would like to be when they grow up and are often rewarded for setting high goals. Peers in lower-class communities also discourage each other from striving for low-paying, low-status jobs, but there is often the implicit acceptance that many of those who strive for high-status jobs will not succeed. A 14 year study by Schiamberg (cited in Irvine, 1990) found that 50% of black boys in middle school aspired to professional or technical occupations, but by adulthood, the figure had dropped to 7%. Fewer white boys aspire to such jobs, but a higher proportion actually obtain them.

After reviewing the enormous literature on both self-esteem and level of aspiration, Cook (1979) concludes, "It does not appear possible on the basis of recent research either to support or discount the assertion that government-enforced segregation lowered the self-esteem and educational aspirations of black children" (p. 425). He discounts the suggestion, however, that the Supreme Court was somehow misled by the researchers at the time. Those who drafted the original statement never suggested that desegregation by itself would have profound effects on levels of aspiration and self-esteem. The statement noted, "It is difficult to disentangle the effects of segregation from the effects of a pattern of disorganization commonly associated with it and reflected in high disease and mortality rates, crime and delinquency, poor housing, disrupted family life and general substandard living conditions" (cited in Cook, 1979, p. 425). Even modest gains, however, depend on a number of conditions, including (1) firm and consistent endorsement by those in authority, (2) the absence of competition between racial groups, (3) equivalent status, and (4) racial contacts that would promote interracial understanding.

Gerard (1983) has examined each of these in detail and has argued that they are rarely met, and as a result, desegregation has not had nearly the impact that was originally intended. He points out that "those in authority" include the entire chain of command from classroom teachers to principals, school boards, and town councils. That chain of command is literally as strong as its weakest link, and any opposition can undermine the commitment of others. The behavior of teachers has the most obvious and most direct effect, but if the town council or school board is unable or unwilling to allocate sufficient funds, it can have a ripple effect that resonates throughout the school system. The quality of education obviously depends on teaching, but it also depends on resources, support staff, and so on. Gerard attributes some of the lack of community support to the "white flight" that occurred as a result of community busing. This has led to a resegregation of many school districts and a subsequent loss of funds.

A second necessary condition for desegregation to work is the *absence of competition among groups.* Numerous social psychological studies have shown that competition produces conflict and aggression, even when groups are distinguished by relatively superficial features (e.g., Sherif & Sherif, 1953). Gerard has pointed out that competition runs deep and pervades every aspect of American society, including the school system. This competition is enhanced by a competitive grading system in which children are clearly marked for success or failure. The strong ingroup and outgroup distinctions that develop within this context help promote interracial tension.

There also appears to be a culture gap between teachers, who tend to be white and middle class, and lower-class minority children. Both white and black teachers tend to underestimate the intelligence of black children, particularly when they use ghetto speech. Black students, particularly boys, are also far more likely to receive criticism and mixed feedback in which praise and criticism occur at the same time (Irvine, 1985, 1986). Because of the high levels of competition within schools and the tendency to develop ingroup-outgroup distinctions, it is not surprising that the second condition, which requires an absence of competition among groups, is rarely met.

The attainment of *equal status* among participants within desegregated schools has been equally difficult to obtain. There appears to be an increasing gap in the level of performance between white and black children as they move through the school system. By the sixth grade, the gap is more than two grade-equivalents wide. Schools also tend to sort children into one of two tracks—one for those who are expected to succeed academically and one for those who are not (Ogbu, 1985). Among those black children who do respond well to school, girls outnumber boys by a ratio of two to one. The tendency to use "magnet" schools for children of high ability, tends to reduce the relative importance of race within these schools but it also takes more talented black students away from the black community (D. Dupree, personal communication, March 1990). The sorting process starts early, so that by the third grade, the worst class in each school has become in effect a detention center to control deviant children and troublemakers, who have little hope of academic success. These children are then drawn away from the schools into the street culture where they adopt various coping strategies to get along.

The final condition for successful desegregation is *interracial contacts* of the kind that would allow students to cut through racial stereotypes and treat each other as unique individuals. The evidence suggests, however, that such stereotypes tend to persist in integrated schools because, to a large extent, self-segregation prevents meaningful contacts between members of different racial groups. Both blacks and whites tend to select their friends and work partners from members of their own race. Even informal groups and seating tend to be divided along racial lines. Integrated schools also achieve a great deal of segregation by using advanced classes in various subjects, where access is based on achievement tests and previous grades. Since blacks score lower on achievement tests they are less likely to be placed in advanced classes (Irvine, 1990). Outside the classroom, children are often bused back to their segregated neighborhoods where interracial contact is even more limited.

Gerard (1983) attributes the failure of desegregation to the *lateral transmission of values hypothesis,* which assumes that classrooms consisting of a small number of minority students and a larger number of white peers would produce a change in values and achievement motivation within the minority group. It was assumed that black students would absorb the values of their white classmates and eventually raise their level of aspiration. The ideal proportion of blacks was assumed to be somewhere between 20% and 40%. Twenty percent provided a critical mass, whereas 40% ensured that whites would still form the majority. These proportions seldom occur, however, and even when they do, self-segregation limits interracial contact. When differences between groups are small, some form of "bootstrapping" does occur, but as the gap increases, there is a tendency for each group to limit their comparisons to members of their own race.

Irvine and Irvine (1983) have described some of the *secondary disruptive effects of desegregation.* Prior to desegregation, black schools were one of the most important institutions within the black community. Following desegregation, they were often dismantled. In many schools, desegregation radically changed the composition of the staff while the student body remained the same. Black teachers and principals, who had previously served as role models, were dismissed or demoted and replaced by new and inexperienced white teachers who served their time and left as soon as possible. Many brought customs that seemed strange and foreign, whereas others were openly hostile to black students. Picott (1976) has noted that the number of black principals in the South declined from more than 2,000 in 1964 to less than 200 in 1973. Most of the 1,800 who lost their jobs became assistant principals, classroom teachers, or administrators of "special projects," with very little real control. Because black schools had previously played such a central role within the black community, the disruptive effects within the school carried over to the community as a whole. Black parents no longer had the same degree of input into the schools, nor could they assume that it would reflect their own interests and values.

All schools have not undergone the same degree of deterioration. Researchers have found that, although some schools with low-income, black students have low levels of achievement, low expectations, and unruly and apathetic students, others—often in the same community—with a similar composition of students have a much better academic climate. What distinguishes effective from ineffective schools is not an optimal balance of white and black students but other factors, such as strong leadership, an emphasis on basic skills, frequent feedback, and a firm conviction that *all* students can learn and it is the responsibility of the school to teach them (Edmonds, 1986). Irvine (1988) has suggested that effective schools also use a strict system of discipline in which there are clear and well-known codes of conduct. They also tend to encourage parental and community participation. The frequent contact among teachers, parents, and members of the community helps to make the school into a strong mutual-support system that discourages aberrant behavior and promotes academic achievement.

The failures rather than the successes of desegregation have been stressed because many people simply assume that desegregation has been tried and it has not worked. They, therefore, conclude that there is little else we can do. Desegregation has not reduced differences between black and white children because of a number of concrete conditions that are in many ways unique to the educational process as it is currently practiced. The competitive nature of the educational process in which schoolchildren work independently on separate problems trying to outdo each other to get a good grade could easily be supplemented by a more cooperative approach in which interracial groups work together. More advanced students could then serve as both teachers, who help other less advanced students, and models. The contact, cooperation, and mutual support under these conditions would help eliminate some of the obstacles that have prevented desegregation from working in the United States. Some alternatives to contemporary education will be discussed in the final chapter.

SELF-HANDICAPPING STRATEGIES

A final word should be devoted to self-handicapping strategies. There is a close link between level of aspiration and self-handicapping because one of the most common self-handicapping strategies is to not try to achieve. Self-handicapping is by no means limited to blacks. It occurs among all segments of society—rich or poor, black or white, male or female—and is probably one of the most important reasons people fail to develop their true potential.

Self-handicapping occurs because attributions for success and failure can be based on either stable, internal factors, such as ability, or other factors, such as external circumstances or effort. A person's recognition that he or she has little ability is so potentially devastating that people sometimes create obstacles which interfere with performance in order to have an alternative explanation for failure. If they manage to succeed in spite of these obstacles, then their perception of their ability is enhanced.

The first study on self-handicapping was carried out by Berglas and Jones (1978). These researchers gave male and female students either soluble or insoluble problems and then told them they had performed very well. They reasoned that a good performance on insoluble tasks would create a sense of uncertainty about the ability to perform well later on. Subjects were then offered a choice between two drugs that would either facilitate or inhibit their next performance. Male subjects who had previously worked on soluble problems generally chose the drug that would improve performance, whereas those, whose experience led them to believe that they would probably not do very well, showed a strong preference for the inhibiting drug. Tucker, Vuchinich, and Sobell (1981) later replicated the study but used alcohol as the dependent measure. These studies suggest that alcohol and drug use might be, at least in part, a form of self-handicapping (see Jones & Berglas, 1978).

The simplest form of self-handicapping and the one that is probably most often used is to simply not try at all. Mark Snyder and his associates (e.g., Frankel & Snyder, 1978; Snyder, Smoller, Strenta, & Frankel, 1981) have suggested that individuals who expect to fail can reduce effort and use this as an explanation for their poor performance. They found that subjects who had previously succeeded but who were uncertain about their ability reduced effort. Not trying is based, at least in part, on the belief that academic ability and effort are inversely related—that is, that smart people do not have to work hard and that those who do are not really smart. This perception that effort and ability are inversely related may not occur outside the academic domain. We take it for granted that talented musicians must practice often to develop their musical ability or that world-class athletes must train hard to reach their full potential. This form of self-handicapping is particularly insidious not only because it is extremely common but because people who do not try and fail seemingly have no one to blame but themselves.

Another form of self-handicapping is the creation of external obstacles that can interfere with performance. Rhoewalt and Davison (1984) found that subjects who had previously performed well but who did not expect to do well on a second trial chose to listen to performance-debilitating music rather than facilitating or neutral music. Subjects who have been led to believe that they might fail on future tasks are also more likely to select tasks that are very difficult (Greenberg, 1983). Failing a difficult task is commonplace and does not suggest low levels of ability. Such failures, like other forms of self-handicapping, provide no information at all about the person's level of ability. People with both high and low levels of ability fail difficult tasks. Other forms of self-handicapping that have been suggested include losing notes before an exam, not getting enough sleep, or becoming overly involved in other activities (Berglas, 1987).

Interestingly, the use of self-handicaps sometimes *improves* performance by reducing anxiety (Brodt & Zimbardo, 1981; Weiner & Sierdo, 1975). When people have an alternative explanation for poor performance, the need to perform well is reduced. Leary (1983), for example, found that socially anxious people who were exposed to distracting noise interacted more successfully than those who were not. Whether or not a handicap facilitates or hinders performance depends on the nature of the task and the type of handicap selected. Reducing effort, for example, would seem to interfere with most tasks.

Researchers have offered several explanations for self-handicapping. First and most obvious is that it represents an attempt to maintain self-esteem where failure is possible (Jones & Berglas, 1978). The people most likely to use self-handicapping strategies are those who have had a mixed history of success and failure and who harbor doubts about their own ability. Instead of trying hard to see how much they can accomplish, some people prefer to create obstacles that provide a ready excuse for poor performance. Arkin and Baumgarden (1985), on the other hand, argue that self-handicapping strategies are a form of impression management, which is used to maintain a positive social image and impress other people. They point out that the majority of studies on self-handicapping use conditions in which subjects are evaluated by others.

In either case, the end result is more or less the same. People who use self-handicapping strategies create obstacles that interfere with performance, so they are never able to accurately assess their true ability and never develop their skills by pushing themselves to the limit. This again shows how naive it is to assume that people try as hard as possible and differences in performance are due to differences in ability. Many people avoid challenge because they have a low self-esteem or a high fear of success or failure. Some do not try to develop their skills because they have low levels of aspiration or achievement motivation. Some avoid tasks because they assume that they have little control over what happens. And some intentionally place obstacles in their path to excuse themselves for poor performance.

The effects of achievement motivation, locus of control, and level of aspiration, however, do not have the same effect on all cognitive tasks. The level of performance also depends on the amount of intrinsic interest. Some people who perform well in one area may do poorly in another. Intrinsic motivation interacts with all the factors discussed previously and helps determine the choice of activities, the way people think and process information, and the development of new cognitive skills.

CHAPTER 6

Intrinsic Motivation

This chapter explores the development of intrinsic motivation by looking at several theories that have concentrated on the topic. This will be followed by a discussion of working conditions in contemporary industrial societies that tend to limit intrinsic motivation, followed by a brief description of domain-specific cognitive skills and the "overjustification effect," whereby external reinforcement may undermine intrinsic motivation. Each section examines intrinsic motivation as a somewhat unique set of acquired drives in which a person derives pleasure from certain activities without any ulterior motive.

Intrinsic motivation is often defined in a purely negative sense as the state of a person who strives toward a particular goal without external reinforcement. When people receive rewards such as money or praise, they are said to be extrinsically motivated. If there is no apparent external reward, then they are believed to be intrinsically motivated. Implicit in the concept is the belief that such tasks are intrinsically interesting and that they are carried out for their own sake.

The concept of intrinsic motivation was developed largely as a reaction to traditional *drive-reduction theories* put forth by Freudians and learning theorists, such as Clark Hull. In Hull's (1943) learning theory, for example, drives are an essential part of the learning process. They activate people and keep them active until the drives are fulfilled or until some other process, such as exhaustion, intervenes. More complex behaviors were explained through the concept of secondary drives, which are learned through association with more basic primary drives, such as hunger, thirst, sleep, and pain avoidance. Affection, for example, was attributed to the fact that nurturing parents provide food and therefore become secondary reinforcers. This association later generalizes to other people who resemble the parents in some way.

It gradually became apparent that drive-reduction theories had serious limitations. Researchers found that animals of all kinds were strongly motivated to explore their environment and manipulate objects without any apparent ulterior motives (Berlyne, 1950; Harlow, 1953). Harlow (1953), for example, showed that young monkeys would learn a complex series of responses to receive a mechanical puzzle and would engage in numerous puzzle-solving trials without any external incentive. Other studies showed that rats would cross an electrified floor to gain access to unfamiliar settings and that human infants would work for long periods on activities that had no apparent external reinforcement. Such behavior could easily be explained in terms of evolutionary theory but could not be explained through drive reduction. The concept of secondary reinforcement was initially used to explain these findings, but as Berlyne (1966) noted, exploration occurs so quickly after birth that it cannot have occurred through association. Such behaviors are also remarkedly resistant to extinction, suggesting that something more basic is responsible.

As Deci (1980) points out, the initial response to such research was to postulate a variety of new drives, such as a manipulation drive (Harlow, 1953), an exploratory drive (Montgomery, 1954) and a sensory drive (Isaac, 1962), but this approach was soon abandoned because the number of drives was growing rapidly and because the new drives did not fit the traditional definition. Drives are usually associated with a state of deprivation that motivates people to seek objects or substances capable of reducing them. The needs to explore, manipulate objects, and maintain a certain level of stimulation do not seem to be based on deprivation at all, and they appear to continue indefinitely as long as stimulating objects are in the environment.

In 1959, Robert White proposed a more general theory of *competence* that attempted to integrate much of this literature. According to White, people and other animals have a basic need to deal effectively with their environment and master and control things around them. This basic need leads to persistence and energizes such behaviors as exploration, attention, thought, and play. The need for competence causes people to seek new challenges rather than waiting for events to happen, and such challenges are intrinsically motivating.

FUNCTIONAL AUTONOMY OF MOTIVES

Gordon Allport (1897–1967) felt that a common feature of all previous theories of motivation was a desire to reduce all motives, no matter how unique and complex, to a limited number of basic biological drives that were shared by everyone and were presumably innate. For Allport (1937), motives and interests were almost infinitely varied and could not be accounted for by any combination of basic biological drives. Human motives change and grow during the course of development, creating more and more diversity as people mature.

Allport felt that any theory of motivation that purports to account for human behavior must recognize this diversity and account for it, and his concept of *functional autonomy* was designed to do just that. The basic principle is that activities initially carried out for other reasons can become self-motivating and sustain themselves without any external or biological reinforcement. Each motive has a definite source derived from either instincts or tissue deprivation but as the individual matures, the bond is broken. Such behaviors begin as a functional attempt to cope with some aspect of the environment, but with sufficient repetition, they become an integral part of a person's character and cease to serve the original function.

Allport (1937) used several examples to illustrate the process, but one of the most colorful is the ex-sailor's craving for the sea. Such a person is likely to have first encountered the ocean in his attempt to earn a living. Later the same person may have become a wealthy banker, and the original need is completely lost; the craving still persists, however, and grows stronger. The longing for the sea has become an end in itself and no longer meets the original needs. A second example is the musician's desire to return to her instrument after a period of separation. The musician may have been originally motivated by harsh criticism of her inability to play. Now she has mastered the instrument and no longer has to worry about such criticism, but she still longs to play. Workmanship is another example. Good workers feel compelled to do a good job even though their security and the respect of others does not depend on high standards. Good workers spend a great deal of unnecessary time making sure everything is precisely right, often loosing time and money as a result. Such needs can become so central that they are valued above everything else.

Allport did not go into a great deal of detail about why certain motives become functionally autonomous, but two factors seem to be involved. First, he claims that there is a strong relationship between interest and ability. People like to do what they can do well. A student who takes a course because it is required or comes at a convenient time of day may become so absorbed with the topic that his or her interest lasts a lifetime. Once an interest has been aroused, it not only creates tension but helps select and direct behavior. People with a particular interest seek out and notice things that others consider irrelevant and recall information more effectively.

Second, people take a strong interest in incompleted activity. Allport drew from the work of Kurt Lewin and his students showing that incomplete tasks create a state of psychological tension which persists until the activity has been completed. Children just learning to walk, talk, or tie their shoelaces take an intrinsic interest in such tasks and engage in them for their own sake. They remember them, return to them, and feel frustrated if they cannot continue. Allport felt that motives always strive for completion. Goal-directed behavior creates a state of unresolved tension that demands a certain degree of "closure."

Some motives such as a desire to learn or perfect a craft are inherently incomplete. A person never learns all there is to know, and craftspeople never develop their skills to perfection. Gifted people strive to develop their ability, even when no external rewards are involved. According to Allport (1937), "An author, a housekeeper, a public speaker, a reformer, a craftsman, a musician, if deprived of their favorite occupation, may 'itch' to return to their work" (p. 322). The pursuit of literature, art, music, and scientific research are all examples of drives that started for other reasons but now have taken on a life of their own. Such drives can become so central that more basic drives, such as eating, sleeping, socializing, and sex, are temporarily deferred until the activity is complete. Although most of the preceding examples have focused on what Allport calls "cardinal traits," many motives can become functionally autonomous, and the specific combination and their relative importance help to make each person unique. There are countless examples of people who get "turned on" to various activities and then pursue them relentlessly without any external incentive. For Allport, the concept of functional autonomy was an attempt to account for concrete impulses that vary from person to person and lie at the root of human behavior.

During the 1960s, Abraham Maslow (1908–1970) developed a similar theory based on his concept of *self-actualization.* Self-actualization can be treated in two different ways—as the ultimate source of motivation in Maslow's hierarchy of needs[1] and as a series of choice points in which the individual selects the more challenging of several behaviors. Because self-actualization is rare, there are a number of misconceptions about what is involved. One of the most common errors is to think of self-actualization in static terms as some sort of destination at which an individual ultimately arrives. From this perspective, self-actualized people need do nothing more than rest on their laurels secure in their knowledge that they are superior to everyone else. For Maslow, self-actualization is not a fixed state of flawless competence, but is instead a never-ending process of developing new skills.

It is useful to see self-actualization as a series of *choice points* in which people must choose between an older, more routine form of behavior and a new more challenging activity often accompanied by uncertainty and the threat of failure. Some of the choices represent major decisions that may permanently alter the course of a person's life. Someone may have to choose, for example, between a career in law or some paralegal profession. A law degree is more demanding, takes longer, and presents more obstacles to success, but it brings more responsibility and more rewards in terms of self-development.

[1] Maslow's (1970) *hierarchy of needs* consists of five basic drives—physiological needs, safety needs, belonging and love needs, the need for recognition and self-esteem, and self-actualization. These needs are typically portrayed as a pyramid with the more basic needs at the bottom and the more complex needs at the top. Higher needs are less essential for sheer survival and can be more easily postponed or delayed indefinitely. Higher needs require more favorable outside conditions, but they also lead to higher states of satisfaction. Lower needs must be fulfilled before higher needs become a source of motivation. The process of development can therefore be seen as a movement up the pyramid in which the fulfillment of lower needs sets the stage for the experience of new needs that then motivate and direct behavior.

Other decisions are much more mundane. Once a student has selected a course of study, he or she must choose between studying hard or pursuing leisure activities. This choice involves countless little decisions such as whether to write a term paper or go to a party. Those who study hard learn more and are more likely to finish their education. Later, people have to make similar decisions, such as whether to pursue a challenging line of research or to stagnate within their profession. Students who fail to study or professionals who choose to stagnate will never develop their potential, not because they lack the ability, but because they fail to push themselves sufficiently hard to develop their skills.

The notion that self-actualized people *push* themselves to the limit is a bit misleading because a great deal of "pull" is associated with these activities as well. Students with an interest in a subject do not necessarily make a sacrifice when they stay home to study. Nor does a research scientist pursuing an interesting line of research. Such activities become intrinsically motivating. Indeed, such activities are often responsible for what Maslow (1970) describes as "peak experiences." These are moments of intense joy in which people are completely absorbed. Peak experiences are the periodic reward for those who have chosen a life of challenge.

JOB SIMPLIFICATION AND ALIENATION

One of the reasons self-actualization occurs so rarely in our society is that the physical and social conditions necessary for achieving it seldom occur. Western society now provides most people with the means for fulfilling their basic biological needs and also offers many a reasonably high degree of security. But meaningful relations with other people and recognition from others are far more difficult to obtain, and for most people, self-actualization is not only an illusive goal but beyond comprehension. Although industrialization has provided a noticeable increase in living standards, it has also created a complex division of labor in which many jobs have become simplified to the point where they are boring and repetitious and offer little intrinsic satisfaction. People's inability to find satisfaction in their work is so commonplace that many simply assume all work is tedious and unfulfilling and that people work only when forced to do so.

The simplification of labor is actually a relatively recent development that began near the start of the industrial revolution. As early as 1776, Adam Smith (1776/1937) noted that 10 men who divided up a complex task could produce much more than the same 10 men in isolation. He described the process of making pins: One man draws out the wire, another straightens it, a third cuts it, a fourth points it, and a fifth grinds the top to receive the head. Making the head requires two or three more distinct operations. Another person puts on the head, and the pin is then whitened and placed into a piece of paper. Smith estimated that 10 men dividing the labor could make up to 48,000 pins a day, whereas one man working alone could produce no more than 20. Although it is widely believed that assembly-line procedures began at the turn of the century, they were actually a fundamental part of the industrial revolution (see Galbraith, 1977).

A century later, Karl Marx (1818–1883) traced the transition from preindustrial to industrial conditions in England and abstracted what he considered to be general economic laws. Feudal society was based primarily on self-sufficient peasant farming, in which workers carried out a variety of tasks. There were also, a large number of craftspeople and guild workers who designed and produced tools, clothing, and furniture. This system, although terribly inefficient by modern standards, was sufficient for the most part to meet the limited needs of a sparsely settled rural population. By the late Middle Ages, however, several changes radically altered the means of production.

The discovery of the New World and the expansion of trade routes to the Orient created new markets for manufactured goods. As these markets grew, feudal modes of production were no longer sufficient to meet the expanding needs. Groups of people were brought together in factories, and the rate of production was increased through a division of labor. The division of labor simplified the process and made the introduction of machines possible. The first industries to develop were the clothing industries, which were stimulated by the increasing demands for wool. This demand for wool led to the Enclosure Movement, which replaced peasants and tenant farmers with sheep and transformed England's preindustrial rural population into a largely urban pool of potential workers.

Competition among factories that produced similar goods meant that goods must be produced as cheaply as possible. The cost of production was reduced by paying each worker as little as possible, increasing the length of the working day, using cheap sources of labor such as women and children, spending as little as possible on the workplace, and introducing machines to replace people. Marx felt that there was an *inherent conflict* between workers and factory owners because workers wanted higher salaries, shorter working days, better working conditions, and full employment; but each of these increased the cost of production. It is not necessary to see factory owners as evil or greedy because they too were victims of economic reality. One factory owner could not unilaterally reduce the working day or pay higher salaries even if he wanted to because this would have increased the cost of production.

According to Marx, changes in working conditions caused four different types of *alienation*—alienation from nature, alienation from work, self-alienation, and alienation from others. Preindustrial people lived in natural conditions. They worked in largely rural environments and transformed raw materials into finished products. The rhythm of labor was tied to the changes of seasons and often included periods of frantic activity punctuated by periods of leisure. Industrial workers at the start of the industrial revolution lived in large cities cut off from nature and spent most of their day in bleak urban factories. The items they worked with were often man-made and unnatural.

Alienation from work was seen as the primary source of alienation leading to both self-alienation and alienation from others. For Marx, people were distinguished from other animals by the fact that they plan activities before they carry them out. Marx (1867/1906) pointed out, "A spider conducts operations that resemble those of a weaver, and a bee puts to shame many an architect in the construction of her cells. But what distinguishes the worst architect from the best of bees is that the architect raises his structure in his imagination before he erects it in reality" (p. 157). For Marx, productive work was one of the most essential aspects of human life. People develop themselves through their work, and it shapes the way they think and relate to others.

Marx felt that one of the most serious problems of industrial capitalism was that it robbed people of the opportunity to carry out productive work and led to alienation. Industrial workers are alienated in two senses. First, they do not own either the raw materials or the finished product; they transform materials but have no control over either the design or the final product. Second, each worker is simply one link in a long production process; the task is simplified and reduced to a few mechanical movements. Industrial work is boring and repetitious and requires little skill or training. One worker can be easily replaced by another, and they can take little pride in their work because it requires so little skill and imagination.[2]

Alienation from work tends to produce *self-alienation.* Craftspeople and professionals can take pride in their work and see it as an integral part of their own self-concept. They can derive pleasure from using their skills and developing their potential. But industrial work has little intrinsic value. It is simply a way to earn a living. Because workers are largely unskilled and interchangeable, their value is determined not by what they do but by what they earn. Workers come to see themselves as commodities and regard others the same way. This leads to *alienation from others* in which people come to be seen as items of exchange that can be bought and sold freely and whose very existence is shaped by conditions outside their control.

[2] Marx was not the first to observe that working conditions help shape the way we think. A century earlier Adam Smith (1776/1937) noted:

> The understanding of the greater part of men are necessarily formed by their environment. The man whose whole life is spent in performing a few simple operations, of which the effects too are, perhaps always the same, or nearly the same, has no occasion to exert his understanding. . . . He naturally loses, therefore, the habit of such exertion and generally becomes as stupid and ignorant as is possible for a human creature to become. (pp. 735–736)

It should be stressed that Marx was describing conditions during an early stage of industrial capitalism when workers labored up to 16 hours each day and earned just enough to keep themselves and their families alive. Working conditions were hazardous and child labor was rampant. Child labor did not begin with capitalism, but the frequent use of machines meant that it was now possible to use children in jobs previously carried out by adults. Industrial accidents were grounds for dismissal, and those who could not find work were at the mercy of charitable agencies or sometimes starved.

The history of the labor movement largely reflects a gradual improvement in working conditions. Child labor laws were established, minimum wage was implemented, working hours were gradually reduced, and companies were forced to provide safer working conditions. Each change faced a great deal of resistance from factory owners who argued that the government should not interfere with business but they kept the situation from deteriorating. Indeed, it is probably one of the greatest ironies in human history that Marx, by showing the "inevitable" consequences of free enterprise, helped save capitalism.

One aspect of early industrialization has been retained, however. There is still an extensive *division of labor* and a strong tendency to break down complex tasks into a few simple operations that can be performed by almost anyone. Job simplification has been retained because it is efficient and lowers the cost of production. During the early 1950s, there was a major change within industrial countries as they moved into what has been called the "postindustrial" period. White-collar workers began to outnumber blue-collar workers for the first time, and this trend has accelerated in recent years. But white-collar work is often subject to the same type of simplification—jobs are reduced to a few mechanical steps that almost anyone can carry out. Thus, many people today feel overqualified for their jobs and find little intrinsic satisfaction.

A factor distinguishing Marx's theory from other theories described previously is that Marx focused on changes occurring *later in life*. For our purposes, the most important question is, Do industrial conditions help shape mental ability, or more precisely, does mental ability decline as a result of job simplification? Several sources of evidence suggest it might. The most direct support comes from the longitudinal research carried out by Kohn and Schooler and their associates. Kohn and Schooler (1978), using various measures of job complexity and intellectual flexibility, studied nearly 700 workers over a 10-year period. They used the term *substantive complexity,* which was defined as the degree that work required thought and independent judgment. They found that, while both substantive complexity and intellectual flexibility were extremely stable over the 10-year period, there was also a reciprocal relationship between the two variables. People with high intellectual flexibility tended to end up in jobs that required a great deal of independent thought, and those in jobs that were not substantively complex became less intellectually flexible. A careful analysis of the data suggested that, although socialization during childhood plays an important role in the development of intellectual flexibility, job complexity and the degree of challenge continue to affect mental ability, independently of the selection process that assigns people to particular jobs and independently of people's effort to change their jobs to make them more compatible with their own needs.

Kohn and Schooler (1978) go on to suggest that, if two people of equal intellectual flexibility enter the work force in jobs that offer different degrees of challenge, the one in the more challenging occupation will become more intellectually flexible and more likely to seek out and obtain jobs requiring more intellectual flexibility. As a result, small differences in substantive complexity early on may have a reverberating effect throughout the life cycle. Because substantive complexity is closely linked to occupational status, socioeconomic differences tend to produce an increasing gap in intellectual flexibility among those in high-status and low-status occupations.

More recently, Miller, Slomczynski, and Kohn (1987) compared young, middle-aged, and older workers in both Poland and the United States. They found that the amount of occupational self-direction, which was based on substantive complexity, degree of supervision, and the amount of repetition in the workplace, tended to affect workers in all three age groups, more or less equally. This suggests that working conditions continue to affect intellectual flexibility throughout life.

These effects are not limited to the workplace, however. People who do intellectually demanding work continue to show more intellectual flexibility outside the workplace (Miller, Schooler, Kohn, & Miller, 1979) and even seek out intellectually demanding activities in their leisure time (Miller & Kohn, 1983). Because their jobs reward initiative and independent judgment, they come to value self-directed activities, both for themselves and their children, and tend to be less authoritarian in their approach to child rearing (Kohn, 1963). Again, these differences tend to have a reverberating effect not just within the life of the person most directly involved but across generations.

It may well be that people's ability to rise to the occasion is based on the job falling within what Vygotsky (1930–35/1978) has described as the "zone of proximal development." Work, like play and formal education, is what Vygotsky calls a "leading activity," in which people develop skills by interacting with other people. If workers come to work with sufficient intellectual flexibility, then challenge and responsibility may stretch their intellectual ability and make them more flexible in their thinking. If they lack intellectual flexibility, then challenges may be too strong and they may quit or seek employment that is more commensurate with their own level of ability. In either case, there is a complex interaction between what people bring to their work and what the job offers in the way of intellectual stimulation.

Indirect support for the role of the environment later in life comes from research that has found changes in "general" intelligence over time. Some of these changes occur rather rapidly. The IQs of schoolchildren, for example, appear to decline during the summer break and recover during the school year (Heyns, 1978; Jencks et al., 1972). The effect is particularly strong among lower-class children, whose summer activities least resemble those found in school. Other changes are much more gradual. Longitudinal research on declines in mental ability later in life have typically found little or no decline among highly educated subjects in intellectually demanding occupations, but among those in less demanding occupations, declines in some areas have been found (e.g., Owens, 1966; Cunningham & Owens, 1983). Both of these lines of research seem to support a *use it or lose it* principle of intelligence. The environment continues to shape mental ability not simply because people develop cognitive skills by adapting to their environment but because without intellectual stimulation, skills previously acquired gradually diminish over time.

Even if conditions improve, many people in our society, through no fault of their own, still find little intrinsic satisfaction or interest in their work. The skills required are often learned in a day or two, and then there is little additional challenge or opportunity for growth. Most people prefer to have a job because it provides a number of tangible and intangible benefits—an income, health insurance, a pension, plus a chance to meet other people and structure the working day—but the job does not provide much intrinsic joy. Alienation is so pervasive in our culture that it is often assumed to be part of the human condition. But boring and repetitive work is not natural. It stems from an overemphasis on job simplification brought on by a forced division of labor. Productive work is intrinsically satisfying, but mechanical work requires just enough concentration to dull the intellect and prevent free use of the imagination.

DOMAIN-SPECIFIC KNOWLEDGE AND COGNITIVE SKILLS

Although it is widely assumed that the ability to perform well within a particular domain is based on a combination of knowledge and more general information-processing skills associated with perception, memory, and problem solving, recent evidence suggests that basic skills associated with information processing are themselves highly dependent on previous experience (e.g., Chi, 1985; Ceci, 1990; Glaser, 1984, 1987). What people know helps determine what they can learn, what they will remember, and their ability or inability to solve new problems. Knowledge within a particular domain is in turn strongly affected by intrinsic motivation. As Ulric Neisser (1967) has pointed out, "We have all known, or been, boys who could remember everything about baseball or fishing but not a bit of history" (p. 288).

Knowledge effects information processing in at least three different ways. First, it helps determine how information is organized. People who are very knowledgeable within a particular domain not only have more knowledge but tend to structure that knowledge differently. They have considerably more specific concepts that draw attention to particular details. This knowledge tends to be organized in terms of higher-order categories and cross-referenced in many different ways. As Chi and Ceci (1987) point out, the critical factor in the development of expertise is the way that facts are integrated and differentiated within the person's knowledge base.

A good deal of research has shown that the amount of knowledge also helps determine how fast and effectively people can process new information. Some of the research on chess masters was described in Chapter 2, and this research has been replicated within the domains of bridge (Charness, 1983), electronics (Egan & Schwartz, 1979), and computer programming (Jeffries, Turner, Polson, & Attwood, 1981; McKeithen, Reitman, Rueter, & Hirtle, 1981). Schneider, Korkel, and Weinert (1990), for example, found that children with lower IQs who knew a great deal about soccer, outperformed their higher IQ classmates on reading, inferential, and memory tasks that included new information about soccer. As Ceci (1990) has pointed out, there is a growing body of evidence that reading, memory, and inferential reasoning are predicted by individual differences in knowledge at least as much, if not more so, than by individual differences in IQ (e.g., Bjorklund & Muir, 1989; Walker, 1987). Such differences depend not just on the amount of knowledge but on the way it is stored. Relevant information about highly interesting activities tends to be well structured and easily accessed.

The amount of previous knowledge also helps determine how new information is used. According to Glaser (1987), information-processing research in the 1960s and 1970s tended to focus on problem solving on "knowledge-lean" tasks that required little previous experience and could be learned in a relatively short time. More recent research, in contrast, tends to focus on "knowledge-rich" tasks that require hundreds and sometimes thousands of hours of previous experience. The latter research makes it possible to compare problem solving in experts and novices and to uncover some striking differences. Problem solving in an unfamiliar area tends to be object oriented and based on actual things and events occurring in the immediate situation. Experts, on the other hand, tend to conceptualize problems in terms of higher-order abstract principles that can be used to organize various aspects of the immediate situation. They have rapid access to a whole body of meaningful material allowing them to go to the heart of a problem and solve it almost by "intuition." Experts also tend to use more self-regulatory, metacognitive processes, such as planning ahead, allocating time, and monitoring and correcting ongoing processes. Such self-regulatory skills tend to be domain specific, although some may cut across several domains or be applied to situations in general.

Intrinsic motivation helps determine not just the acquisition of knowledge but the development of specific skills needed to perform well within a particular domain. Ceci (1990) gives several examples of studies that have looked into the mismatch between "general" intelligence, as measured by IQ tests, and intellectual performance on specific tasks. Dorner and Kreuzig (1983), for example, had subjects assume the role of a city manager for a mythical town called "Lohhausen." Over one thousand different variables were identified and the researchers devised a procedure for differentiating nine different levels of problem-solving ability, but they have consistently failed to find *any* relationship between the level of problem solving and IQ.

Ceci and Liker (1986) have carried out a similar study of expert race handicappers. They began by identifying two groups of men who attended horse races nearly every day of their adult lives on the basis of their ability to predict odds at post time. One group, whom they called "experts," seemed to use complex mathematical procedures to identify horses who were under- and overrated. The experts not only considered more information, but they combined it in a nonadditive way. To predict the speed of a horse in the last quarter mile of a race, experts seemed to use a seven-way interaction model, which took into account not just the closing speed but the distance from the rail. Within groups (expert and nonexpert), there was no relationship between IQ and the ability to calculate speed and, therefore, place bets. Between groups, experts with low IQs consistently outperformed nonexperts with high IQs. What is more, there was no relationship between the ability to use this interaction model and scores on the Arithmetic subscale of the Wechsler Adult Intelligence Scale.

A more direct effect of intrinsic motivation on learning and problem-solving ability can be seen in a study by Bronfenbrenner and Ceci (cited in Ceci & Nightingale, 1990). When these researchers asked 10-year-old children to estimate the distance that large or small, black or white circles and triangles would move on a screen, their performance was abysmal and there was little improvement even after 750 trials. When the task was transformed into a video game, however, and the triangles and circles became birds and butterflies, they learned to perform much better. Ceci and Nightingale (1990) point out that, if the experiment had been limited to the dull disembedded laboratory task, it would have led to unnecessarily low estimations of children's ability to consider multiple causes when they reason.

Findings such as these seriously challenge the notion that "general" intelligence cuts across specific domains and increases performance on a wide range of tasks. What passes for general intelligence may be based on the acquisition of a large number of partially overlapping skills acquired within specific domains and later transferred from situation to situation. Ceci and Nightingale (1990) suggest that processes and strategies that cut across domains in adulthood often have their childhood origins within particular domains and cannot initially be deployed effectively outside those domains. Some skills may not transfer, however, and whether they do or do not depends, at least in part, on intrinsic interest in the task itself. Even such basic skills as the speed or depth and breadth of mental processing seem to be strongly affected by practice and familiarity (see Ceci, 1990). The effects of knowledge on memory and information processing will be discussed in more detail in Chapters 9 and 10.

PROMOTING AND UNDERMINING INTRINSIC MOTIVATION

Although traditional learning theories assume that learning is based on reinforcement, reinforcement sometimes interferes with learning and performance. The undermining of intrinsic motivation through the use of external reinforcement has become known as the *overjustification effect.* Some of the earliest studies of the overjustification effect were carried out by Deci (1971, 1972), but the study most frequently cited is probably that of Lepper, Greene, and Nisbett (1973).

Lepper, Greene, and Nisbett (1973) provided nursery school children with an opportunity to play with felt-tip pens and drawing materials and then observed their behavior. Once baselines had been established, the children were assigned to one of three conditions. The first group was told in advance that they would receive a Good Player Certificate if they played with the materials. A second group received the same certificate at the end of the session but without forewarning, whereas the third group neither expected nor received the reward. One week later the children were given an opportunity to play with the same materials during a regular classroom period, where they were secretly observed through a one-way mirror. Children in the expected-reward condition spent half as much time drawing as the other two groups.

The overjustification effect appears to be quite robust and has been found in a large number of studies (see Condry, 1977; Lepper & Greene, 1978; or Ross, 1976, for reviews). According to Condry (1977), people who are given external rewards for performing intrinsically interesting tasks,

> choose easier tasks, are less efficient in using the information available to solve problems, and tend to be answer-oriented and more illogical in their problem-solving strategies. They seem to work harder and produce more activity, but the activity is of a lower quality, contains more errors, and is more stereotyped and less creative than the work of comparable non-rewarded subjects working on the same problem. Finally . . . subjects are less likely to return to a task they at one time considered interesting after being rewarded to do it. (p. 472)

In short, the use of external rewards to reinforce intrinsically interesting tasks not only decreases motivation but alters the way people approach and process information.

Rewards do not always undermine intrinsic motivation, however. As Bandura (1986) has pointed out, self-motivation often requires skills that are initially learned through external incentives. Many activities are initially tedious and uninteresting but become interesting after the person has reached a certain level of competence (cf. Allport, 1937). Bandura (1986) states:

> Children are not born innately interested in singing operatic arias, playing tubas, solving mathematical equations, writing sonnets, or propelling shotputs through the air. But with appropriate learning experiences, almost any activity, however silly it may appear to many observers, can become imbued with consuming significance. (p. 241)

Deci (1975, 1980) suggests that rewards serve *two functions.* They can be used either to inform people about their level of ability or to shape their behavior. When control is made more salient, the information function is reduced and so is intrinsic motivation. Rewards that are not based on the level of performance may be seen as an attempt to control behavior, and lead to lower levels of activity after they are removed. People simply complete the tasks as soon as possible to achieve the reward but lose interest in them when rewards fail to occur. Extrinsic reinforcement can reduce intrinsic motivation in two ways—by altering the perceived cause of behavior or by lowering a person's assessment of their own ability.

Following Heider (1958), Deci argues that the level of intrinsic motivation is based in part on the *perceived locus of causality.* People see the locus of causality as internal when there are no external incentives. An external incentive may come to be perceived as the cause of behavior, particularly when it is not closely tied to the level of performance. The overjustification effect is therefore partially based on the attributions people make about *why* a behavior occurs, and these judgments help determine both their level of interest and their level of performance.

External reinforcement also provides information about people's level of ability, which helps determine the level of *perceived competence.* In general, verbal or monetary rewards that inform people they are becoming better at a particular task increase the level of intrinsic interest, whereas those that signal a decline in ability lower the level of intrinsic motivation. A decline in interest may therefore be due either to a shift in the perceived locus of control or to a perceived drop in ability. Deci, Cascio, and Krusell (1975) suggest that there may be gender differences in the way people interpret events. They found that praise had a different effect on men and women. Praise strengthened intrinsic motivation in men (presumably by increasing their assessment of their own ability), but it decreased intrinsic motivation in women (presumably by changing the perceived locus of control).

In short, the use of external incentives can be effective under certain conditions, especially when they are directly linked to the level of performance and provide feedback that someone is doing well. External reinforcement can also stimulate interest in tasks that have little intrinsic interest to begin with. A study by Johnson et al. (1984) provides a good example. These researchers gave IQ tests to black inner-city children and white children from the middle class. Half the children in each group were given tokens, which could be exchanged for toys, for each correct response. Black children who were given tokens scored 13 points higher than those who received no reward, but tokens had no effect on the performance of white children, probably because they were intrinsically motivated to begin with. Studies like this one suggest that people do not always perform well even when they have a sufficient level of ability. The use of external incentives often assumes, however, that the activities themselves will eventually become intrinsically satisfying. Intrinsic motivation can be increased by providing people with challenging tasks and not undermining their sense of self-determination through the unnecessary or excessive use of extrinsic reinforcement.

Although achievement motivation, locus of control, level of aspiration, and intrinsic motivation have been treated separately in this text, there is probably more overlap among these constructs than the literature would suggest. High levels of aspiration and achievement motivation promote the development of skills in areas that later become intrinsically interesting. Achievement motivation itself, like intrinsic motivation, is often domain specific and it appears to operate only when people believe they can personally control what happens to them. We know from Dweck's research that intrinsic motivation is one of the variables that distinguishes mastery-oriented and helpless children. When confronted with a difficult task or the possibility of failure, mastery-oriented children see the task as a challenge, become more interested, and increase effort. Helpless children, on the other hand, devalue the task and reduce effort. Each of these motivational constructs has cognitive consequences and helps determine not just the degree of interest but the development of cognitive skills.

PART THREE

Cognitive Skills

CHAPTER 7

Language and Thought

Any discussion of the relationship between language and thought must begin with a brief overview of language itself. The first section focuses on the nature and development of language. It is followed by a brief review of the more limited topic of concept formation and then by a discussion of the relationship between language and thought. The chapter ends with an examination of the linguistic relativity hypothesis and some of the limited research that has been conducted. The critical question is whether linguistic groups differ in the way they perceive and process information and, if so, whether these differences really matter.

Although language and thought develop later than other more basic psychological processes, such as perception and memory, there are a number of reasons to begin Part Three with a discussion of language. First, the development of language radically alters perception and memory. More basic skills are transformed once language has been acquired, and these changes help determine what is noticed and what is not, how information is processed, and what is recalled. Second, the study of the relationship between language and thought and the concept of linguistic relativity has a relatively long history. It can, therefore, serve as a general introduction to more specific treatments of the social origins of perception, memory, and problem solving. Those who support the doctrine of linguistic relativity claim that thought is shaped by language and that linguistic differences produce different ways of thinking.

Language and thought are so closely related that many people fail to notice the connection. Many psychologists and philosophers have difficulty dealing with concepts such as consciousness and thought, because they take their own conscious experience as the starting point. When the French philosopher, René Descartes (1596–1650), tried to derive the one indisputable claim on which he could base his philosophy, he ended up with "I think, therefore I am." What Descartes overlooked and what many people have failed to notice since is that thought itself has a starting point and it too is derived from other activities, such as the use of mental images and language.

Two centuries later, when the German psychologist Wilhelm Wundt (1832–1920) tried to build a new discipline on verbal reports of conscious experience, the relationship between language and thought was equally elusive. Wundt, who is commonly regarded as the founder of modern psychology, spent the first part of his career developing an experimental psychology based on introspection and the latter part developing a social psychology or "Volkerpsychologie" that examined the social aspects of mental processes, as expressed in cultural products such as language, customs, and myths. Wundt's belief that language was a product of consciousness (rather than the other way around) obscured the connection between language and thought and led him to develop two distinct disciplines—experimental psychology and social psychology—each equally important yet disconnected.

Since then, several theories have been developed by Lev Vygotsky (1886–1934), Ludwig Wittgenstein (1889–1951), and George Herbert Mead (1863–1931) that stress the intimate relationship between language and thought and the social origins of each. These men, working without knowledge of each other's work, have put together comprehensive social theories that have radically altered the way these concepts are treated. Each suggests that concepts develop during conversation. Once people have begun to communicate, they can begin to direct their conversation at themselves. This capacity for inner speech then becomes the basis for verbal thinking. Vygotsky (and to a lesser extent, Wittgenstein and Mead) explicitly distinguishes between thought and language, but for each, language is an important tool for solving problems and communicating ideas that becomes progressively more important as children mature (see Collier, Minton, & Reynolds, 1991).

LANGUAGE DEVELOPMENT

When one first begins to examine languages, what is most striking is the almost unlimited diversity. Even closely related languages, such as French and English, differ greatly. The English word "dog," for example, is "chien" or "chienne" in French, depending on the gender. In English, adjectives come before the nouns they modify (e.g., the big white house), but, in French, they can come either before or after (e.g., la grande maison blanche) and they are, again, marked for gender. The English language gains precision through an extensive vocabulary of more than a million words, whereas the French rely on a more precise grammar in which common English ambiguities simply cannot occur. Closer examination, however, reveals certain similarities. All languages are made up of basic sounds, or phonemes, which are then formed into words. Words are put together according to grammatical rules to form sentences and so on.

Phonemes are the smallest, most elementary units of speech. They include basic sounds associated with vowels and consonants plus other sounds such as *ch, sh,* and *th.* The number and types of phonemes vary from language to language. Some have as few as 18, others have as many as 85. English has about 45. Even familiar languages, such as French and German, have phonemes that do not occur in English, and unless these are learned relatively early, an English speaker may have difficulty distinguishing certain words or pronouncing them properly.

Phonemes are put together in different combinations to form *morphemes,* which are the basic meaningful units. There are two classes of morphemes—freestanding and bound. Freestanding morphemes are complete words that have an independent meaning. Bound morphemes are prefixes and suffixes that are attached to other words and alter the meaning. The word "nonconformists," for example, includes five morphemes—*non, con, form, ist,* and *s.* There are over 100,000 morphemes in the English language and these can be put together to form over a million English words.

Words are formed into sentences according to rules of *grammar.* The formal grammatical theories developed by linguists attempt to describe the underlying regularities within a language. These hypotheses describe the structure of language and attempt to account for the unconscious processes used more or less automatically by native speakers. They attempt to describe what speakers must "know" to speak and understand a language properly.

Language development follows a similar path from simple to complex, beginning with phonemes and ending up with a complex, unconscious understanding of grammar itself. Newborn infants in a delivery room turn their head toward the source of a sound (Wertheimer, 1961). Infants 2 to 3 weeks of age can discriminate speech from other sounds. By 1 or 2 months, infants can discriminate between consonants, such as *b* and *p* (Eimas, Siqueland, Jusczyk, & Vigorito, 1971) and begin to make cooing sounds themselves. They start to babble several months later. Babbling involves the prolonged production of a wide variety of speech sounds. It is not based on imitation for it includes phonemes of various languages not heard in the home. Deaf children babble as well, even though they cannot hear.

The babble of children from different cultures is indistinguishable at 5 months (e.g., Nakazima, 1962), but differences begin to occur at 6 months, and by 10 months the native language can be identified. These differences are based on the gradual elimination of sounds not used as phonemes in the native language and a gradual development of new sounds that are used. By 1 year, children begin to babble in ways that resemble sentences and produce a gibberish that often sounds like a statement or a question.

At about the same time, they begin to produce their *first words.* The first words are usually single-syllable words that are easy to pronounce and begin with consonants such as p, m, d, t, or b. They may be real words that refer to important people, such as "da" or "ma," or personal constructs, such as "baba" for bottle or "buff" for the family dog. These personal constructs may be barely recognizable but they are used consistently, and family members quickly learn to understand them. Many of the first words refer to things that move or can be played with rather than equally common stationary objects, such as tables and chairs (Nelson, 1973). Animal terms tend to be common, even among urban children, whose parents point them out in books (de Villers & de Villers, 1979). These single-syllable words carry a variety of meanings, and the same word can be used as a statement, a question, or a request. Gestures, intonation, and context are used to provide additional information and make the meaning clear. A child may say "milk," for example, and hold a cup up at the same time or may point to a white liquid and say milk with a quizzical look and a rising intonation. Words at this stage are often overextended and refer to objects that are either perceptually similar (Clark, 1973) or have a similar function (Nelson, 1973).

At about 1½ years of age, children begin to put words together into *two-word sentences.* These phrases typically include a noun and a modifier, such as "big shoe," "big boy," "more milk," or "milk allgone," and the words tend to occur in more or less the same order. This invariant order is the first sign that children are abstracting underlying regularities based on rules of grammar. Later three- and four-word sentences are formed, but the speech of young children remains "telegraphic" because articles and other inessential features are omitted. Later, these words are added, and the speech of older children becomes more and more like that of adults.

A second sign of grammatical development occurs between 2 and 3 years of age when children began to make *systematic mistakes* in plural and past tense. They begin to create words that they have never heard, such as "feets," "mouses," "goed" and "ranned." These mistakes occur because many common English words do not follow regular forms. Children abstract regular rules, such as forming plurals by adding "s" or creating past tense by adding "ed," and then apply them to cases where English makes an exception. They *overregularize* the rules of grammar and apply them incorrectly. Similar patterns have been noted in other languages as well (e.g., Slobin, 1970).

Virtually all children learn grammar quickly and seemingly effortlessly before the age of five. The speed and accuracy of language development have led some to claim that humans have an innate propensity or predisposition to acquire language (e.g., Chomsky, 1965, 1975, 1980, 1986). Those who take this position draw support from three basic sources: (1) the fact that language development occurs in children without direct instruction; (2) the difficulty most adults have in learning a second language; and (3) the presence of universal features.

Parents teach language only indirectly—by serving as models, correcting mistakes, and speaking down to their children. Although traditional learning theories claimed that language is learned through rewards and punishment, neither positive nor negative feedback about performance appears to be a sufficient explanation. Brown and Hanlon (1970) examined parents' tendency to reward children's speech and found that the crucial variable was the accuracy or truth of a statement. Parents are not particularly bothered by ungrammatical or incomplete sentences, and they make little systematic effort to correct them. Parents correct statements that are factually incorrect and to a lesser extent mispronunciations, "naughty" words, and overregularized plural and verb forms (Brown, 1973). Nelson (1973) has found that when mothers systematically correct mispronunciations and reward good ones, their children learn vocabulary more slowly than other children. Chomsky argues that language development is underdetermined because there is insufficient stimulation to account for what children actually learn.

Many adults find it difficult to acquire a second language. They do not find it easy to distinguish and articulate sounds that do not occur in their native language. A *critical period* for the acquisition of phonemes appears to occur early in life. Unless some exposure occurs during this period, adults may have difficulty learning to speak a second language.

A third line of evidence supporting the innateness hypothesis is that all languages share certain features. Chomsky feels that a *universal grammar* is genetically built in and that all languages are simply variations of this common pattern. The universal grammar restricts the range of possible alternatives and instinctively directs children away from alternatives that do not fit. Brown (1986) adopts a similar position and argues that linguistic universals "exist because of cognitive constraints, because the mind of the human species creates a kind of abstract blueprint that all languages must fit" (p. 468).

Others disagree. Some have challenged the claim that language is underdetermined. Although parents do not seem to reinforce or shape language development directly, they serve as models and help simplify and structure the learning environment. Young children learn language as well as other skills by interacting with people who are older and more competent than themselves. Parents and older children "talk down" to young children and help them deal effectively with situations until they are capable of doing things by themselves. They work within what Vygotsky (1930–1935/1978) has described as the "zone of proximal development," which is highly tailored to meet the needs of children and is systematically altered as children mature. When adults speak to young children, they use short, simple, and well-formed sentences with exaggerated intonation. This form of speaking has been called "motherese," although it is certainly not limited to mothers. Even 4-year-old children modify their speech when talking to younger children (Shatz & Gelman, 1973).

Once children master a new activity, the caretaker "raises the ante," so to speak, and place additional demands on the growing child. This constant adjustment and readjustment forms the basis for Bruner's (1983, 1985) notion of *scaffolding,* which makes it possible for children to learn complex activities, such as language, in incremental steps. Young children spend virtually their entire waking day playing and experimenting with language, and parents often provide a continuous running commentary on what is happening. Parents name objects, ask questions, and expose their children to new concepts and ideas. Parents and other adults also tend to repeat young children's well-formed sentences and "recast" errors, particularly when there is only one mistake. If a child says "I *breaked* the cup," for example, the parent may say "So, you *broke* the cup, did you?" Such recasting both alerts the child about the source of error and serves as a source of clarification about what the child means. After hearing thousands of simplified sentences, it may be that children simply discover that nouns precede verbs and later abstract other features in much the same way.

The difficulty that adults have in learning a second language also cannot stand as irrefutable proof that language is innate. Many of the problems that adults have when they attempt to learn a second language stem from the inability to hear and reproduce sounds; there may well be a critical period for the development of basic sounds. English-speaking adults, for example, find it difficult to discriminate phonetic contrasts that occur in Czech but not in English. Trehub (1976) has found, however, that babies from English-speaking homes have little difficulty with such contrasts. This suggests that infants are born with the ability to distinguish a wide range of sounds but that they gradually lose this ability if the contrasts do not occur in their native language. Other studies have also shown that infants can discriminate speech sounds that do not occur in their mother tongue, even when their parents cannot (e.g., Lasky, Syrdal-Lasky, & Klien, 1975; Werker, Gilbert, Humphrey, & Tees, 1981).

Other problems emerge because many adults treat the acquisition of a second language as a kind of game in which they consciously master the rules of grammar and use them to construct sentences. This top-down approach is the precise opposite of the way children learn a first language—that is, from the bottom up without any explicit knowledge of language structure. Problems also occur because the first language interferes with the development of a new language. New speakers often try to translate the old language into the new by finding precise equivalents and using a similar word order. Finally, there is the problem of practice. Learning a second language is difficult because it is periodic and lacks the total immersion of first-language acquisition.

The notion that there is an innate program for the development of a universal grammar has also been challenged. Hebb, Lambert, and Tucker (1973), for example, claim that linguistic universals occur not because of innate programs but because of the nature and structure of reality. When people deal with each other or the world, their actions often take the form of an agent acting on objects, thus creating nouns and verb forms. Objects differ in shape, size, and color, so adjectives or some equivalent adjective form attached to nouns are necessary to distinguish them. Every language appears to have ways for expressing height, length, and distance and directions such as up-down, front-back, right-left, and so on. All languages have ways of distinguishing past, present, and future events, singular and plural forms, and ideas and their negation. It is not unreasonable to assume that the need to deal with these aspects of reality has forced people throughout the world to create similar grammars to deal with common problems.

The reason this debate is important is because the absence of innate programs for language development means there may be substantial differences among languages, over and above the common features needed to adapt to similar environments. Because there is a close connection between language and thought, these differences may mean that linguistic differences help shape the way people perceive and process information. This is the basic argument behind the concept of *linguistic relativity* and it will be discussed later in more detail.

If there are universal constraints for language development based on innate programs, they must be at a fairly high level of abstraction. Bruner (1985) argues that Chomsky's innate Language Acquisition Device (LAD) must be supported by a partially innate Language Acquisition Support System (LASS), which is based on social contact. There are, however, enormous cultural and even subcultural differences in the way children learn language (see Rogoff, 1990). The intimate face-to-face interaction which is the focus of research in Western cultures is by no means universal. In many cultures, young children spend a great deal of time on their mothers' backs, where they can listen in on conversations and learn language indirectly. In some societies, children are not encouraged to speak to adults as conversational partners. They are expected to speak when spoken to.

In cultures where adult-children conversation is limited, older children begin to play a greater role in language development. They also simplify their speech so that it is easily understood. John-Steiner and Tatter (1983) have pointed out that language development tends to occur within a context in which the primary goal is communication. To communicate, people *must* tailor their speech to those who are being addressed. This tendency to simplify speech for young children seems to be a natural part of communication and occurs among virtually all peoples who have been observed.

Language development occurs on two fronts simultaneously. While children are learning grammatical rules and regularities, they are also developing and expanding their vocabulary. This second type of learning based on concept formation also depends on abstracting underlying regularities and features, but the focus is on individual words rather than sentences and phrases.

CONCEPT FORMATION

Although there is a close connection between concepts and words, the two are not the same. Human infants and other animals make categorical distinctions without the use of language. Many people believe that concepts are learned first and labels are acquired only after prelinguistic categories have been formed. The development of language, however, radically alters and accelerates concept formation. It facilitates communication, allows people to incorporate new information, and ultimately makes it possible to explicitly define the features that distinguish one concept from another.

Concepts are the basic units of thought that allow people to place objects within general categories and structure an otherwise bewildering array of information. Verbal concepts include terms such as "dog," "cat," "red," "white," "pretty," "ugly," and so on that cover a number of different objects or qualities that are considered more or less the same. They allow people to organize information into manageable units and transfer information from one situation to another. If we can label something, the label allows us to fix the object in our mind and relate it to other objects and events.

It was once thought that people learn concepts by providing a dictionary definition that specifies the necessary and sufficient features. This *classical theory* or feature-based approach was challenged by the Austrian philosopher, Ludwig Wittgenstein (1889–1951). Wittgenstein (1953) argued that words need not refer to distinct objects or even common features of ideal objects. Instead, they are tied together by a *family resemblance,* which is a set of overlapping features shared by some members of a set but not by others. Each instance has one or more features in common with other instances that distinguishes them from other concepts. For Wittgenstein, words are tools and like other tools—hammers, pliers, saws, screwdrivers, rulers, glue-pots, and squares—they come in many different forms and serve a variety of different functions.

The concept of family resemblance can be seen more clearly in Wittgenstein's (1953) example of *games.* There is no common feature characteristic of all games. Some are amusing, some are intensely competitive. Some involve team work, others are played alone. Some, such as chess, have intricate rules; whereas others, such as a child bouncing a ball against a wall, have no rules at all. Instead of a single underlying feature, there is a network of "overlapping and criss-crossing" characteristics. They are like shared family traits, where each person resembles other family members but in different ways. A daughter has her father's features but her mother's dark hair and complexion. A son looks faintly like his uncle around the mouth and chin but has his grandfather's eyes and temperament. What is true for games and physical features is equally true for more abstract concepts such as "goodness" and "beauty." Like fibers of a rope, concepts do not depend on any single feature but on a combination of numerous overlapping and disconnected threads. The openness of concepts means that there will always be a certain looseness about them. Concepts that pick out common properties must be sustained as new cases are added. For Wittgenstein, concepts originate not in the mind of an individual but in shared social activity and the practical world of everyday experience. Each individual acquires concepts by incorporating new cases into concepts previously learned from others.

Wittgenstein's concept of family resemblance was extended by Eleanor Rosch (1973, 1978), who argues that some concepts are easier to learn because they are more "natural." *Natural categories* have two characteristics. They are more basic and prototypical. According to Neisser (1987), basic categories have members that look alike and serve similar functions. Rosch and her associates have shown that a composite picture of a basic category, such as a "chair," created by superimposing different examples over each other, still looks like a chair. Chairs also tend to elicit a similar response—"You first turn your back and then lower your rump to meet the seat" (Neisser, 1987, p. 14).

Basic categories also have a *medium degree of inclusiveness or abstraction.* Rosch distinguishes three levels of abstraction—superordinate, basic, and subordinate. Superordinate concepts, such as fruit, clothing, vehicles and so on, are the most abstract and contain the greatest number of specific cases. Basic concepts, such as apples and oranges, pants and shirts, cars and trucks, are moderately inclusive and have an intermediate level of abstraction. Subordinate concepts—McIntosh or Delicious apples, blue jeans or dress pants, BMWs or Mercedes-Benzes—are the most specific and contain the fewest number of specific instances.

Rosch (1973, 1978) also argues that most categories are encoded in terms of *prototypes.* Some members of a category seem to be particularly good examples, whereas others are marginal. Prototypes are examples that best illustrate a specific concept. Prototypes have the greatest amount of "family resemblance" and are in a sense at the center of a category. They have more features in common with other examples of the same concept and are more sharply differentiated from other similar concepts. Some dogs, such as Golden Retrievers, for example, are more prototypical. They seem to be more like "real dogs" than Pekingese or Chihuahuas. Robins and sparrows are prototypical birds, apples and oranges are prototypical fruit and prototypes exist for most categories. The notion of what is typical varies from person to person, and there are cultural differences as well. While robins and eagles are typical birds in the United States, swans and peacocks seem more typical from a Chinese perspective.

Concepts and concept formation vary with age and experience. Words that children and adults use to describe objects do not necessarily mean the same thing. Children's early concepts are frequently *overextended* or used too liberally to apply to objects not covered by the concept. A young child may use the word "dog," for example, to refer to any four-legged creature—cats, cows, horses, and so on. Later, other words are learned and the concept of dog is used more correctly. Caretakers are often aware of these differences and tolerate words that are technically wrong. A whale, for example, may be described as a "fish," because the distinction is too subtle for a child to grasp.

Katherine Nelson (1974, 1986) argues that children initially name objects and understand events not in terms of physical properties but on the basis of their *function.* Balls, for example, can be bounced, thrown, caught, or kicked. Functional relationships are tied together by scripts, which are general patterns of behavior characteristic of stereotypical situations, such as eating breakfast or going to school. Nelson has found that objects that can be acted on, such as dogs and balls, are learned earlier than those that are not. Infants as young as 9 months of age can sort objects on the basis of function, distinguishing, for example, objects associated with eating, such as plates, pans, cups, and spoons, from those associated with grooming.

Even very young children can and do sort objects on the basis of *physical characteristics,* however. In one study, Oliver and Hornsby (1966) gave children 33 familiar objects such as tools, silverware, fruits, and toys. They found that 4-year-olds tended to group objects with similar physical characteristics. Bananas and pears, for example, were placed together because they were both yellow. Children between 6 and 9 years tended to sort items on the basis of function. Bananas and pears were placed together because they could both be eaten. Older children and adults tend to use more abstract categories, such as fruit, to link objects at the intermediate level.

Although functional and perceptual features are important aspects of early concept formation, concepts are also organized in terms of higher order categories (Neisser, 1987). Concepts such as insects and frogs, for example, are linked by an implicit understanding that they are animals. Being an animal implies certain needs, such as the need to keep warm and find food. It is unlikely that each of these features is encoded separately as each concept is learned. The fact that a bug or a frog needs to eat and be kept warm is simply irrelevant for the most part, unless someone wants to keep them as a pet and is faced with the problem of finding food and shelter.

This suggests that much of our knowledge of specific objects is constructed on the spur of the moment, to deal with specific problems and situations. It tends to be what Barsalou (1987) has called *context-dependent.* We draw from a large body of interrelated information and construct concepts on particular occasions. The features that stand out may vary from situation to situation. With repeated exposure, certain features begin to stand out and become context independent. Most people cannot think of the concept *skunk,* for example, without thinking of *smell* at the same time. The association eventually becomes obligatory, and there is no way of preventing it. A feature becomes context independent only after it has occurred many times, and the degree of exposure varies from person to person. Context-independent information eventually forms the common core that defines a specific concept and remains relatively stable over time.

Such experiences eventually lead to a qualitative shift in the nature of concepts (Keil, 1987). Whereas young children understand concepts in terms of typical features, people gradually learn to define concepts in terms of a small number of necessary and sufficient conditions. This *characteristic-to-defining shift* occurs on a domain-by-domain basis that varies from person to person. This suggests that unrefined, childlike categories for unfamiliar concepts are common even among adults. The shift is strongly influenced by the need to share knowledge. People attempt to reach a common ground by tacitly agreeing on what is important and unimportant. Initial concepts with fuzzy boundaries become more compact and tightly defined. Interestingly, this shift corresponds to the distinction between classical and the more recent theories of concept formation. As people gain experience, their ideas change so that they correspond more closely with the classical notion of defining concepts in terms of necessary and sufficient conditions.

Knowledge within a particular domain changes concepts in other ways. Basic-level concepts, for example, change with experience. Rosch, Mervis, Gray, Johnson, and Boyes-Braem (1976) use the example of an experienced airplane mechanic for whom types of airplanes have become the basic category. As people gain knowledge, the features that stand out change, so experts can make fine distinctions that other people overlook. Novices judge by appearance; experts use more sophisticated cognitive models (Neisser, 1987). Anyone who really understands the concept of "mammal" will recognize that mammals do not become fish simply because they now live in water or because they have undergone evolutionary adaptations that have transformed the limbs previously used for locomotion on land. Concept formation thus occurs in both a top-down and bottom-up direction at the same time. The number of levels and degree of differentiation ultimately depend on how important a concept is for a particular person or a particular culture.

THOUGHT AND LANGUAGE

Thinking can be defined as the manipulation of *symbols* to process information or solve problems. Words are symbols, but they are not the only kind. A symbol is anything that can stand for or refer to something else. Other symbols include mental images, sign language, and the abstract symbols used in logic and mathematics. Albert Einstein claimed that his thinking was initially inspired by complex kinesthetic or bodily feelings from which he derived his notions of time and space. Testimonials of this sort demonstrate the complexity of thought and the variations that occur in the way people think. Different symbols have various advantages and disadvantages, and they can, therefore, be used selectively to deal with different types of problems.

The French philosopher Jean-Paul Sartre (1905–1980) has provided a particularly good description of *mental images.* Sartre (1936/1963) began with the age-old question, How do we know what we experience is real and not just a figment of our imagination? He argued that the question cannot be answered in this form and that a more appropriate question is, How do mental images and perceptions differ? What distinguishes one from the other? He identified four distinct features.

First, mental images are more vague. They are crude representations based on memory that lack the concrete detail of perceived events. This is why many people who can draw well with live models or real scenes find it difficult to draw the same items by imagining them. A second closely related feature is that we discover in images only what we already know. If I try to recall the face of a friend, I may be able to conjure up a rough copy of the original face, complete with eyes, ears, nose, and hair, but the image lacks detail. If I can study the same face, on the other hand, I may discover all sorts of additional features that I had failed to notice previously. Reality is always richer than our recollection of it, and it is possible to gain new knowledge by returning to it again and again.

A third feature includes the faint recognition that our mental images are not real and simply substitute for the real thing. In fact, they are often used precisely for this reason. If I miss someone and long to see her I may imagine her to keep in touch. If this sense of substitution is lost, as it is during some forms of mental illness, then our mental images can seem hauntingly real.

The fourth feature is that we can manipulate mental images but not perception. We can alter perception by selectively focusing on different aspects of the environment, but we cannot alter objects or move them around. We can alter objects through imagination, however, or even create objects that are not real. We can imagine walking to a door, for example, or even through it. It is this aspect of mental images that makes thought possible. By manipulating mental images, we can plan activities, set goals, imagine alternatives, and try out behaviors in our imagination before we commit ourselves in real life.

Certain problems, such as counting the number of windows on the back of one's house, lend themselves to mental images, and they are used more or less automatically. Mental images also serve as a powerful method for aiding memory and recall (see Chapter 9). But there are also certain inherent limitations in the use of mental images. One of the most important is that they tend to be limited to concrete objects or objects that can be visualized. Although it is possible to represent abstract concepts, such as love or peace, through symbols such as a stylized heart or a dove carrying an olive branch, such images fail to capture the complexity and nuances of the concepts themselves. It is also difficult to communicate with images, and therefore, those who think in terms of mental images are stuck with the difficult task of translating mental images into words. Problem solving in early childhood is tied to concrete objects, but abstract concepts become more important as children mature.

People can also use abstract symbols to think, but this is a relatively late development. The advantage of abstract symbols is that they can be used to develop formulas which are always true. Wittgenstein (1922/1961) has stressed that logical statements are unconditionally true because they are *tautologies*. The law of the excluded middle, for example, which states either p or not p, is true because the negation of a true statement is always false and therefore one or the other must be true. The same is true in mathematics. The statement $2 + 2 = 4$ is always true because it is the property of $1 + 1 + 1 + 1$ that it can be rewritten as $(1 + 1) + (1 + 1)$ or $(1 + 1 + 1 + 1)$—that is, $2 + 2$ and 4. The advantage of such statements is that they are always true, and it makes no difference whether the formulas are used for adding apples, bananas, or guided missiles. But abstract symbols are also difficult to understand and people usually start with numerous concrete examples until they discover the underlying regularity.

Mental images and abstract symbols both have certain inherent weaknesses. Mental images tend to be too concrete, whereas abstract symbols are too abstract. Only language has the ability to incorporate both types of concepts within a single system. People can use language to think concretely (e.g., that dog is big), abstractly (e.g., what is true is good) or in some combination of the two (e.g., that dog is good). Because of this, verbal thinking begins to replace "visual" thinking as people mature, and there is a corresponding decline in the use of mental images. Thought becomes more and more tied to language, and it eventually becomes the dominant mode of thinking.

The development of verbal thinking can perhaps best be understood by looking at the work of Lev Vygotsky. Vygotsky (1934/1962) began by making a distinction between language and thought. He felt that they developed differently and had different genetic roots. Early, more rudimentary forms of thinking occur in both young children and other animals. Köhler's (1925) research on insight learning in chimpanzees, for example, showed that they were capable of solving complex problems if all the elements necessary for a solution were visually present. When presented with a problem such as how to obtain a banana suspended from the roof of their cage, Köhler's chimps would look around and notice that there were boxes available. They would then step back, survey the situation and hit on the solution—that is, stack the boxes and use them like a staircase or a ladder. The term "insight" was carefully chosen so as to convey the visual nature of the solution.

Language development occurs somewhat later than thought, through a process described previously. Vygotsky regarded language as a shared social activity and vehemently denied Piaget's contention that children begin with egocentric speech directed primarily at themselves and only later learn to communicate. The development of language for Piaget involves a process of gradual socialization in which the intimate and personal monologues of young children are eventually replaced by dialogues based on mutual understanding. Vygotsky argued that egocentric speech was not the source of communication, nor was it a temporary phenomenon that gradually disappeared as children mature. He saw it as the start of a new type of process based on self-regulation.

For Vygotsky (1934/1962), the primary function of language is communication, and children's first attempts to speak include a strong desire to be understood. Initially, there is a close link between words and the objects they describe, so that the name applied to an object is seen as an important characteristic or essential feature. But a transformation takes place during play. Play is an important source of motivation, but it also helps disassociate words and what they stand for. Children gradually learn that one object can stand for something else. A stick can become a "horse," for example, or a doll can become a "baby." This change in reference leads to the realization that words have an independent status and can be used to solve problems without actually manipulating the environment.[1]

Little by little, children learn to use speech to plan and direct their own activity, and they begin to speak to themselves when they are confronted with a problem or an obstacle. Initially, speech accompanies behavior, but gradually, it moves more and more forward and begins to precede it. It now becomes possible to talk about activities before they occur—to plan and carry out activities verbally before completing them.[2] Young children continue to speak out loud, even when others are not present. Only later do they learn to use language to plan activities and solve problems without actually speaking. For Vygotsky (1930–1935/1978), the decisive moment in intellectual development *"occurs when speech and practical activity, two previously completely independent lines of development, converge"* (p. 34, italics in original).

Vygotsky argued that the "egocentric" speech that Piaget had found in preschool children was actually a transition stage between communication with others and silent conversation aimed primarily at oneself. The proper sequence, according to Vygotsky was not egocentric speech followed by communication but external speech, egocentric speech, and then inner speech. Egocentric speech is the bridge that links communication and thought. It does not disappear, as Piaget claimed, it merely goes "underground."

[1] Peter Hobson (1990) has recently argued that the decisive moment occurs when the infant recognizes that its perspective and that of its caretaker are different. The realization that other people see objects from different points of view, which occurs around 9 months of age, gives objects their independent status and is a necessary precursor to language. Those who lack the ability to take other people's perspective, such as autistic children, often fail to develop language or develop bizarre forms of language in which whole passages are repeated verbatim without any apparent grasp of the meaning of individual words.

[2] Vygotsky (1930–1935/1978) points out that a similar pattern occurs when children identify their own drawings. Young children can label their drawings only after they have been drawn, whereas older children can plan their drawings in advance.

Social speech and inner speech eventually splinter off and serve two separate functions. Social speech becomes more refined and continues to be used primarily for communication, whereas inner speech gradually becomes the dominant mode of thinking. Although social speech and inner speech serve different functions, inner speech is still speech and it is, therefore, based on social interaction. The inner speech of adults has the same function as egocentric speech in children. It is used to plan and direct activities and process and retrieve information.

Inner speech, like egocentric speech, is condensed and abbreviated and it becomes even more abbreviated over time until it becomes virtually incomprehensible to an outside observer. Vygotsky argued that "even if we could record inner speech on a phonograph it would be condensed, fragmentary, disconnected, unrecognizable, and incomprehensible in comparison to external speech" (cited in Wertsch, 1985, p. 173). One form of abbreviation is based on what Vygotsky (1934/1962) described as the *tendency toward predication,* where the subject is deleted and the predicate is stressed. Bruner (1985) considers this tendency very important. He traces the distinction to the Prague School of linguists who used the term "subject" (or topic) to describe what is consciously shared when two speakers converse and reserved the term "predicate" for that which is new. What is already known serves as the background for what is new. One of the reasons inner speech would be largely incomprehensible to an outside observer is that it omits what is obvious and more or less taken for granted.

Inner speech is also characterized by a preponderance of *sense over meaning.* Meaning refers more or less to the dictionary definition, whereas sense depends on the context. Vygotsky uses the example of the fable, "The Dragonfly and the Ant." The ant who is busy putting food away for the winter warns the dragonfly that he had better do the same. When the dragonfly refuses, the ant concludes by saying, "Go and dance!" The words "go" and "dance" have rather precise meanings but in this context they imply both "enjoy yourself" and "die." Vygotsky (1934/1962) points out, "A word in a context means both more and less than the same word in isolation: more, because it acquires new content; less, because its meaning is limited and narrowed" (p. 146).

Inner speech resembles a conversation with an extremely sympathetic listener. Vygotsky uses the example of the characters Kitty and Levin in Leo Tolstoy's *Anna Karenina,* who communicate by using the first letter in words. Levin writes W y a: i c n b, d y m t o n and is astonished when Kitty correctly translates it as "When you answered: it can not be, did you mean then or never?" Kitty responds by writing I c n a o t, which means "I could not answer otherwise then," and then s t y m f a f w h—"so that you might forget and forgive what happened." And the dialogue continues. What makes the example so powerful is that it is based on a similar incident in Tolstoy's own life when he declared his love for the woman who would later become his wife.

Such understanding between people is rare, but speech aimed primarily at the inner self can be extremely abbreviated. It can include all kinds of assumptions that are merely taken for granted. The translation from inner speech to communication may be quite difficult in some cases. We cannot make ourselves understood simply by stating what we think privately. We must go into more detail, define our terms, and often use numerous concrete examples. Even this may not be sufficient. Communication is doubly difficult because words do not fully capture the underlying sense and because people often have subtle differences in their definitions for the same terms. As Waltz (1988) has pointed out:

> Words are not in any case carriers of complete meaning but are instead more like index terms or cues that a speaker uses to induce a listener to extract shared memories. . . . In this sense language may be like the game of charades: the speaker transmits relatively little, and the listener generates understanding through the synthesis of the memory items evoked by the speaker's clues. (p. 197)

There are probably a great many extremely gifted thinkers who cannot communicate effectively and this problem is particularly pronounced among independent thinkers who work out their ideas in isolation. Without communication, our thoughts merely go with us to the grave.

Once thought develops, all other psychological processes are altered as a consequence. Vygotsky (1934/1962) made a distinction between higher and lower mental processes. Lower processes, such as sensation, perception, attention, and will, have a biological origin and are similar in other species, whereas higher processes are culturally acquired and uniquely human. Lower processes continue to function after higher ones have been acquired, but they are modified and restructured. After language has developed, it is used to direct attention, organize perception, establish goals, and guide behavior. Children who previously solved problems impulsively without thinking can now pause, reflect, and view their situation from a "higher ground" (Bruner, 1985).

Vygotsky felt that the focus on lower processes that was characteristic of behaviorism made it totally inadequate as a description of human psychology. All higher processes are mediated processes, and language is the principal tool used to guide and direct them. Verbal thinking contains concepts and generalizations previously absent. Because higher processes depend on language, they are socially acquired and may vary from culture to culture.

LINGUISTIC RELATIVITY

Although elements of the linguistic relativity hypothesis had been introduced previously, the modern concept of linguistic relativity was introduced by Edward Sapir (1884–1936) and developed by his student Benjamin Lee Whorf (1897–1941). According to Sapir (1921), for communication to occur, people must use broad general concepts that tie specific events to others. Because a similar process occurs when we speak to ourselves, thought contain both more and less than the events they describe and reflect individual differences in native languages. This theory, which has become known as the Sapir-Whorf hypothesis, suggests that languages have been developed over the years to meet the unique needs of different populations and cultures. According to this view, which is similar to Wittgenstein's (1953) and Vygotsky's (1934/1962), language is not used merely to express ideas. Language is used to classify and structure the stream of immediate experience, and therefore, variations in language lead to differences in the way people perceive and understand reality.

The most obvious differences are variations in vocabulary. Languages vary in the number of terms and the degree of differentiation. Eskimos, for example, have many different terms for ice and snow. The Hanunoo of the Philippines have separate terms for more than 92 varieties of rice, and Arabs have over 6,000 terms associated with the camel—one for each year of the camel's life, one for each month of pregnancy, one for different cuts of meat, and so on. When objects are important, many different words are used to make numerous fine distinctions. It is possible to translate terms from one language to another but only in a very roundabout way by adding adjectives and modifiers. Common terms that are used habitually in one language may not occur at all in others. Such terms facilitate communication, perception, and recall.

Linguistic differences are not limited to vocabulary, however. Differences also occur at the level of grammar. The most celebrated (and misrepresented) example is the Hopi language, described by Whorf (1956). The Hopi language contains no grammatical forms that refer directly to time as in our past, present, and future tense. Instead, it includes references to both space and time that allow Hopi speakers to think in terms of both concepts simultaneously. Whorf argues that the English language conceals a metaphysics based on the notion of a static three-dimensional space and a perpetually flowing and uniform one-dimensional time—two utterly separate and disconnected aspects of reality. The Hopi's ability to mark space and time simultaneously leads to greater precision and actually comes closer to contemporary theories of space and time based on the theory of relativity.

Hopi speakers also tend to use verbs where English speakers use nouns. They therefore turn English statements about things into propositions about events. Our tendency to use nouns and verbs causes us to polarize the world into objects and events, but nature itself is not polarized. In Hopi, words such as "lightening," "waves," "flame," "meteor," and "pulsation" are all verbs because they refer to events of brief duration.

The tendency to use nouns rather than verbs in English may be partially responsible for the reification of events, which is so common in psychology.

Fingerette (1969), for example, has pointed out that the Freudian concept of the unconscious is frequently misunderstood because it is treated as an object rather than a process. If the unconscious is seen as a failure or a disinclination to bring material into awareness through verbal description, then it becomes much easier to understand. People are conscious of events when they focus on them and maintain a running dialogue and are unconscious of them when they do not.

Similar problems have occurred for concepts such as intelligence and memory. The notion that memory is a thing—a kind of storehouse where previous events are stored—has sent researchers scurrying to find a physical location within the brain. A more dynamic approach, on the other hand, that treats memory as a process of reconstruction (e.g., Neisser, 1967, 1988) helps eliminate many of these problems and creates a better model of how memories are processed and retrieved. The belief that intelligence is a thing quite naturally causes many psychologists and laypeople to view it as relatively stable and resistant to change. If it can be seen as a set of skills acquired during socialization, then some of the difficulties based on the current conception can be eliminated. John Dewey (1922) felt that the concept of "intelligence" was more appropriately described as an *adverb*. People solve problems, behave, and process information "intelligently" or "unintelligently," depending on their previous experience or their familiarity with the situation. Our tendency to objectify events by treating them as nouns has caused many people to miss the dynamic quality of concepts such as repression, memory, and intelligence.

Linguistic differences also occur at more abstract levels. English, for example, tends to use spatial metaphors for things other than space. Spatial metaphors are used to describe time (long, short, and so on), intensity (high, low), and other concepts as well. Spatial metaphors are used so habitually that we seldom notice them. They even occur in the realm of ideas, as the following quote illustrates:

> I "grasp" the "thread" of another's arguments, but if its "level" is "over my head" my attention may "wander" and "lose touch" with the "drift" of it, so that when he "comes" to his "point" we differ "widely," our "views" being indeed so "far apart" that the "things" he says "appear" "much" too arbitrary, or even "a lot" of nonsense. (Whorf, 1956, p. 146)

Although such metaphors are common in English, Whorf claimed that they were totally absent in Hopi speech.

Whorf (1956) has pointed out that the revolutionary changes that have occurred in science during the past century are all based on "new ways of TALKING about facts. It is the USE OF LANGUAGE UPON DATA that is central to the scientific process" (p. 220, uppercase in the original). New concepts have been developed or redefined and old ways of speaking and describing data have been abandoned. According to Whorf, many of these changes are difficult for laypeople to understand, partly because they fly in the face of linguistic conventions.

Although the concept of linguistic relativity is intuitively appealing to some people, it has not received a great deal of empirical support. Many of the more interesting aspects of Whorf's theory have never been adequately tested because they are difficult to operationalize and assess. Whorf's own observations of the Hopi language were based on interviews with an elderly Hopi man living in New York City (A. Neisser, 1983), and they may not be entirely accurate. Few psychologists are sufficiently bilingual in radically different languages to study the problem, and when such attempts are made (e.g., Bloom, 1981), they can usually be faulted on some grounds. Although linguistic relativity seems to be an important area of research, it is not an easy area to investigate and the amount of actual research is quite limited.

There has never been any support, for example, for the claim that the high frequency of nouns in English and other Western languages leads to reification. The concept of reification was developed by the German philosopher Karl Marx, so it is certainly not limited to English. All we know is that nouns are frequently used in cases where verbs or even adverbs might be more appropriate. It is equally plausible that the high frequency of nouns is based on our tendency to reify events or that both are due to some common third factor.

There has been some research on the relationship between grammar and perception. An early study of the Navajo language by Carroll and Casagrande (1958) was carried out to see if their language forced them to pay more attention to shapes. When Navajo speakers use verbs for handling, they must indicate whether the objects being handled are long, short, rigid, or flexible by attaching an appropriate suffix. Carroll and Casagrande reasoned that this causes them to pay more attention to shapes and use them more often to classify objects. When they compared English and Navajo speakers, however, they found no difference.

Most of the research that has been done focuses on differences in *vocabulary.* An early study by Brown and Lenneberg (1954) suggested that the ability to recognize and remember colors was based on the presence or absence of appropriate terms, but more systematic work by Eleanor Rosch (previously Heider) showed that this is not the case. Rosch compared members of the Dani tribe in New Guinea, who have only two terms for color, with American college students (Heider & Oliver, 1972). Each subject was shown a single colored chip and was asked to identify it 30 seconds later from a set of forty colored chips. She reasoned that if the Sapir-Whorf hypothesis was correct, then the Dani would be severely handicapped because colors with the same name would be more easily confused. No differences were found, however. She suggests that the numerous color terms in English is based on a need to distinguish a large number of commercial products that are artificially colored, but the absence of color terms does not affect people's perception of color itself (Rosch, 1974).

Her work is supported by research carried out by Berlin and Kay (1969). These researchers prepared a chart with 320 small squares representing various colors and shades. They then asked native speakers in 12 different languages to point to the best examples of colors in their language. With the exception of those with only two terms for colors, the choices were all more or less the same. The Navajo term "lichi" referred to the same color as the Japanese "aka," the Eskimo "ampaluktak," and the English "red." The boundaries around the colors varied, however. Those with more color terms selected a smaller number of colored chips as examples of the same color. If a language had only two terms, they always referred to light and dark. If a third term was used, it was almost invariably red. The fourth term corresponded to yellow, green, or blue. And so on.

Although language does not radically alter the direct perception of concrete objects, it may determine the way people perceive and organize more abstract qualities. Some of the most convincing evidence occurs in the area of person perception. Numerous studies have shown that schemata for such traits as masculinity and femininity, dependence and independence, introversion and extraversion alter the way people process and recall information. Cantor and Mischel (1977), for example, showed subjects a number of traits that either fit or did not fit a prototype, such as energetic for extraverts. They found that subjects recalled more traits that fit the prototype, and when they were later asked to recognize the traits that had been presented, they also falsely recalled seeing traits that usually accompany it.

Hoffman, Lau, and Johnson (1986) carried out a study designed specifically to test the linguistic relativity hypothesis. They had English-speaking and bilingual subjects read a description of four people who matched two English or two Chinese prototypes. One English prototype described an "artistic type," who had artistic interests and skills and an artistic temperament and cognitive style, and who lived an unconventional bohemian lifestyle. A second English prototype described a "liberal," who was tolerant and open-minded and had progressive, left-wing attitudes and humanitarian and people-oriented values. The Chinese prototypes described a constellation of values and traits that are quite common among the Chinese but have no precise equivalents in English. One was captured by the term *snì gù,* which describes a person who is worldly, experienced, socially skillful, devoted to family, and somewhat reserved. A second term described a person who is very knowledgeable and skilled in a wide variety of areas but reluctant to display these talents unless it is absolutely necessary and is inconspicuous to the point of being ignored.

They found that English and bilingual subjects given the descriptions in English recognized *fewer* items associated with the English prototypes, whereas bilingual subjects given the target descriptions in Chinese were less able to recognize items associated with the Chinese prototypes. Both groups were also less confident in their recognition of previous items associated with the prototype described in their own language and were more likely to identify new items and infer traits that were consistent with the prototype but did not appear in the original description.

These researchers suggest that ready access to an appropriate label makes people less sensitive to details. The common terms summarize a variety of traits and form a stereotype of what people are probably like if such labels are applied. Those without a label, in contrast, must attend more closely to new information and are more accurate at distinguishing what did and did not occur. They contrast their own research with the research on color terms and conclude that color concepts are the exception to the rule. In those cases where natural objects or characteristics can be labeled with a brief verbal description, these labels are likely to be quite useful and aid recall. For more complex concepts, however, verbal labels tend to cover a variety of traits that are not present but implied.

The abstract labels that people use to describe each other seem to influence the way we categorize and subsequently recall information. As Wittgenstein (1953) noted, more complex concepts are open-ended and frequently change. Concepts such as masculinity and femininity have changed radically over the years and other concepts, such as liberal and conservative, are also undergoing fundamental revisions. When new objects are developed or new distinctions are made, then additional terms are either borrowed or invented. In a highly technical society, people within each profession have a number of unique and highly specialized terms that they use to communicate and facilitate the flow of ideas.

Language can also shape information processing in some rather subtle ways. There appears to be a substantial correlation between people's ability to pronounce numbers quickly and their digit-span memory for numbers, which may in turn help to determine their mathematical ability. Welsh children, for example, have a poorer digit-span memory and do more poorly in arithmetic, because it takes longer to count in Welsh than in English (Ellis & Hennelly, 1980). The added time apparently interferes with memory and lowers performance on math tasks. Chinese speakers of Cantonese, in contrast, count more quickly than English speakers and, therefore, have a superior digit-span memory.

There thus appears to be some limited support for a weak version of linguistic relativity. The strongest evidence seems to be based on the use of abstract terms to describe other people. The presence of labels decreases the need to study objects and may limit access to information by limiting attention. People with labels make more distinctions than those without labels. A person who has taken the time to learn the names of various trees, for example, can walk through a forest, recognizing an oak here, a pine there, and a beech tree way in the distance. These are based on differences within languages rather than differences between languages, but similar differences may occur across languages because different peoples have different needs and live in different environments. There is a complex interplay between perception and concept formation. Concept formation is based on perception, and differences between concepts are based on actual differences in the real world. Once a concept has been formed, on the other hand, it may provide feedback and alter people's perception of objects and events. The presence of a label makes identification easy, and it decreases the need to study objects and events thoroughly. People tend to be cognitively lazy, and this laziness is particularly pronounced when it comes to prejudices and stereotypes in person perception. People may not notice subtle aspects of behavior that are literally right before their eyes, and they are, therefore, denied the feedback necessary to correct their preconceptions.

Even though the effect of language on perception is limited, it may play a more pronounced role in *memory*. Once we leave the world of direct perception, words become one of our few links with the past. We reconstruct our memories using language, and these reconstructions seldom precisely capture all the original details. Inaccuracies in memory are based on the inability to go back in time and verify the original experience. In the absence of such verification, our conceptual awareness of previous events may be all that remains. It is extremely difficult to recognize the inherent limitations imposed by language because it is virtually impossible to step outside language and view it objectively. There are no independent observation points, and there is no mechanism for describing language without the use of words. What is known, however, suggests that thought is shaped and constrained by linguistic conventions. These differences help determine the way we approach problems, recall information, and (possibly) perceive the world.

CHAPTER 8

Perceptual Skills

This chapter focuses on subtle aspects of perceptual development that may produce differences in the way people perceive and process information. The main point is that our experiences and, in particular, our social experiences help determine what we notice and what we fail to notice, our ability or inability to concentrate for extended periods of time, and other aspects of perception. The chapter begins with a brief overview of three ecological approaches—those of James J. Gibson, Eleanor J. Gibson, and Ulric Neisser. This is followed by more specific discussions of the way experience helps determine both selective and sustained attention. The chapter ends with a brief overview of perceptual differences based on differences in "cognitive styles."

Although perception is often described in information-processing terms, where a passive receiver sorts through a meaningless array of disconnected stimuli, a better starting point is provided by the ecological approach of James J. Gibson (1966, 1979). According to Gibson, information about the physical world is provided by the structure of ambient light, air vibrations, and chemical reactions which can be picked up by an active organism moving about the environment. The observer is immersed in a sea of highly structured physical energy that conveys the layout of the land and the physical properties of the environment. If a terrestrial surface is nearly horizontal, relatively flat, sufficiently extended, and rigid, it provides support and can be walked on. Each of these properties—horizontalness, flatness, extension, and rigidity—is a physical property and can be directly perceived. According to Gibson (1979), "*Tools, food, shelter, mates, and amiable animals are distinguished from poisons, fires, weapons and hostile animals by their shapes, colors, textures, and deformations*" (p. 232, italics in the original).

By anchoring perception in the physical world, Gibson overcomes many of the dichotomies that have plagued the study of perception—the distinctions between internal and external, subjective and objective, and so on. This is a particularly good starting point for social psychologists because it helps explain how people experience the *same* perceptual world. People actively explore a highly structured and meaningful environment, picking up information about objects and events. There is no need to specify a mechanism by which isolated bits and pieces of information are recombined into a meaningful whole, since the stimuli are meaningful to begin with. As Gibson points out, the secrets of nature are not understood by breaking a code. Perception is a matter of discovering what the environment is really like and adapting to it (Neisser, 1976a). This means that when several people observe the same event, they may not actually see the same "thing." A cliff with a rock face looks different to an artist, a climber, and a child.

ECOLOGICAL THEORIES OF PERCEPTUAL DEVELOPMENT

While James Gibson attempted to build a general model of perception that stresses the importance of information in the environment, others such as his wife, Eleanor Gibson (1969, 1988; Gibson & Spelke, 1983), have focused primarily on changes in perception that occur from maturation and perceptual learning. Gibson contrasts her own position (and that of her husband) with various forms of "enrichment theories." Enrichment theories draw a distinction between bare and meaningless elementary sensations and the more organized and meaningful perception of the same events. They assume that raw sensations are picked up from the environment and these are then interpreted and integrated during the flow of information processing. Sensations are organized by internal processes based on schemata or in terms of inferences or hypotheses about the world. In each case, something else is *added* to the raw sensations to make them meaningful. Enrichment theories tend to stress the privacy of each individual's subjective experience and imply that perception is frequently idiosyncratic and inaccurate.

Gibson (1969) suggests that, instead of viewing sensations as bare and impoverished bits and pieces of information that must be supplemented by internal processes, researchers should consider the opposite. The physical environment is rich in potential information that can be picked up by a sensitive and exploring observer. There is always more information than a person can take in, and therefore, perception tends to be selective. People actively seek meaningful information that can be used to distinguish objects or guide behavior, and they come to ignore features and characteristics which are not relevant. Skilled and unskilled observers differ, not because the former adds something new to the stimulus array but because they detect more features and higher order structures.

Perceptual development consists of an increased ability to extract information from the environment, and it depends on both maturation and learning. Perception becomes more precise, more differentiated, and more refined as people gain the ability to detect properties and features not previously noticed. The recognition of distinctive features depends on discovering properties by contrasting objects that differ in some way. It occurs quite early and is present to some extent at birth.

Perceptual development occurs on several fronts at the same time. Things in the physical world are initially poorly differentiated, but they become more differentiated with experience. This process has four aspects. First, the attention of young children is "captured" by things and events, but they gradually develop the capacity to voluntarily direct their attention. Young infants tend to become fixated on objects and remain fixated until habituation or a scene change occurs. Throughout early childhood, physical properties of objects, such as size, shape, and color, determine what is noticed and what is not, but as they mature, children gradually become more able to selectively focus on important aspects of the environment. Periods of fixation decrease, and visual exploration becomes more common.

Second, perception proceeds from unsystematic to systematic search. There is some evidence of systematic search even in newborns, but it is quite limited. Studies of eye movements in infants and older children show that older children are more likely to explore objects thoroughly. Because young children are unsystematic in the way they explore their environment, they frequently have difficulty identifying two similar objects or distinguishing objects which are different.

Third, children become more selective in their pickup of information. As they mature, they learn to identify larger patterns on the basis of distinctive features and become progressively more able to selectively attend to single aspects of complex situations. Perception of events occurs more easily, and this allows people to focus on smaller and more subtle details. This process continues throughout life but varies with experience. Chase and Simon (1973), for example, found that chess masters are more able to identify larger patterns and recognize larger meaningful chunks of information. Gibson suggests that the difference between skilled and unskilled perceivers is not due to the addition of new elements but to the detection of higher order structures and features that are imperceptible to naive observers.

Finally, children become progressively better at ignoring irrelevant details. As they gain experience, they gradually learn to focus on the most meaningful aspects of information and ignore everything else. This makes it possible to pick up information and make distinctions more quickly. Perception becomes more economical because people learn to attend to a minimal set of features necessary for recognizing objects and making distinctions. Perception thus becomes richer and more selective at the same time—richer because it is more structured and detailed, more selective because nonessential features are no longer noticed.

There is a complex interplay between perception and language in Gibson's theory that can modify perception to a certain extent. Perception occurs without language in young infants and other animals. The perception of objects and their features occurs prior to language and forms the basis for the perception of categories and concept formation. The acquisition of words, however, alters perception in several ways. Words serve as an economical way of guiding and directing attention. Labels allow people to compare current objects with those previously experienced and may increase the efficiency of memory. Labels also facilitate instruction and make it possible for people to communicate and point out critical features and higher order structures. Distinctive features that differentiate one object from another and link objects in terms of their common function must be noticed before they can be labeled, but once concepts are learned, they can exert a selective role in guiding and directing perception.

Ulric Neisser's (1976a) *perceptual cycle model* is similar. Although Neisser has been a close friend of both James J. and Eleanor Gibson and dedicated his book *Cognition and Reality* to them, he felt their theory did not go far enough because it failed to specify the perceiver's contribution to the perceptual process. According to Neisser, the perception of objects depends on both the structure of the environment and the presence of cognitive structures, or *anticipatory schemata,* that direct attention and help determine what is perceived. Anticipatory schemata prepare the person to accept certain information and reject others. Information that fits the existing schemata is readily incorporated, but schemata are also modified and corrected during the course of experience and the cycle continues.

Neisser (1976a) points out that perception is the place where cognition and reality meet and an adequate adjustment to the outside world depends on accurate perception. He feels that current models of perception tend to glorify the perceiver, who is said to process, transform, recognize, and assimilate what would otherwise be a meaningless array of disconnected stimuli. He disagrees and points out that perception, like evolution, is a matter of discovering what the environment is really like and adapting to it. Schemata are derived from interaction with the external world, and they are tuned or adjusted during interaction. Unsophisticated observers have yet to develop complex schemata and tend to focus on relatively superficial features, but people gain experience by interacting with objects and these experiences lead to a more accurate and sophisticated view of the external world. Neisser (1976a) argues that, if the perceptual cycle model is correct, then there could never be a time when infants did not have schemata. Some crude innate perceptual schemata are necessary for perception to occur at all. These undergo periodic revision and updating by a process similar to what Piaget has described as "accommodation" and, out of this, emerges the complex perceptual awareness characteristic of experienced adults.

SELECTIVE ATTENTION

One of the hallmarks of perception is the ability to selectively attend to only a small part of the total potential information and focus on what seems most essential. This capacity to selectively attend occurs quite early. Newborns turn their head in the direction of a human voice. Within days, babies can distinguish their mother's facial expression, odor, and voice. A week-old baby, placed between two gauze nursing pads, will generally turn to the one from its mother's bra (MacFarlane, 1975). Children between two and three months of age spend more time gazing at a drawing of a human face than at a pattern of a bull's eye but they gaze more at the pattern with a bull's eye, which resembles the human eye, than at a solid disk (Fantz, 1961). In short, there is considerable evidence that infants are biologically equipped to attend to important features of their physical environment.

Selective attention becomes progressively more tuned to the external environment as children mature. Even newborns scan visual stimuli but in a very limited way. Young infants tend to fixate on only one corner of a simple geometric figure (Salapatek, 1968) or a single facial feature (Maurer & Salapatek, 1976). As mentioned previously, attention is initially drawn or "captured" by certain aspects of the stimulus array but the ability to voluntarily direct attention improves with age (Gibson, 1969). This tendency is based on both maturation and perceptual learning within a particular domain. Older children and even adults may become fixated and unable to scan complex or unfamiliar objects systematically. Thomas (1968), for example, has shown that, when adults were given a difficult scanning task, such as inspecting X rays for signs of pathologies, those with little experience suffered some of the same defects as children. They failed to scan the X rays exhaustively or focus on the most significant features. Large areas of the X rays were unexplored.

Research on infant perception has recently advanced considerably because of the development of new techniques that are more sensitive to shifts in attention. One of these is based on a tendency for attention to *habituate* as objects become more familiar. Piaget (1952) has suggested that infants are most attentive to moderately new things that can be assimilated but also require some degree of accommodation, and this view has been developed by Jerome Kagan (1971) into what he calls the *discrepancy principle.* Children after the second or third month of age will look at, listen to, or touch things that are moderately different from objects previously experienced.

Kagan (1971) suggests that children begin to form *schemata* at this time, and it is the similarity of new objects to previous objects that determines attention. These schemata for objects are not an exact copy, but a representation that contains some of the original features. Slightly later, children begin to organize information into categories in which they treat different objects as more or less the same. Cohen and Strauss (1979), for example, gave 30-week-old infants pictures of either a single female face or a series of female faces and then introduced a new female face. Those who had seen the series of faces showed less interest in the new face. Such habituation shows a decline in interest due to a lack of novelty, and similar habituation has been found in even younger children.[1] Younger and Cohen (1983), for example, found a similar pattern for infants as young as three to four months using simple categories that varied along a single dimension.

By 9 or 10 months, children begin to organize information in terms of *prototypes,* which then help determine their response to new events. Roberts and Horowitz (1986), for example, showed 9-month-old infants examples of birds and found that they habituated more quickly to prototypical examples not previously seen, such as robins, sparrows, and blue jays. There was no evidence of habituation for atypical examples, such as turkeys, ostriches, or chickens.

Kagan (1971) found *social-class differences* in young children's ability to attend to information. He found no relationship between social class and children's attentiveness to discrepant information at 4 months of age, but by 8 months, a slight positive relationship began to occur, and by one year, infants with better educated parents were significantly more attentive. Children from middle-class homes showed longer "fixation times" or looked longer when they looked at a picture and these differences increased during the second year. Kagan attributed these differences to the greater variety of stimulation occurring in middle-class homes. Similar results have been reported in an unpublished study cited by Kagan (1971). Tulkin observed upper-middle-class and working-class mothers and their 10-month-old daughters at home and under laboratory conditions. Upper-middle-class mothers generally had more face-to-face contact and spoke to their daughters more often. When these girls were observed in the lab, middle-class infants were more likely to switch their attention from old to new objects and were more attentive to taped samples of speech. When a child was placed between her mother and a stranger and a tape recorder played the voice of her mother or the stranger, middle-class infant girls looked at the person whose voice was being played. This tendency to match the voice with the person did not occur among working-class infants and it suggests that middle-class girls as young as 10 months have already developed the tendency to compare schemata and experience. Again, the differences were attributed to the greater variety of stimulation in middle-class homes.[2]

Attention is often limited to one event at a time (cf. Neisser, 1976a). Studies of dichotic listening, for example, have shown that, when people are

[1] Habituation is a technique commonly used to measure an infant's ability to recognize familiar and unfamiliar stimuli. As previously mentioned, infants attend to novel stimuli but habituate when stimuli are repeatedly shown. If a new stimulus is introduced that resembles the original one in some way, the degree of attention can be used as an indication that the infant recognizes the similarity. The use of these procedures has radically altered the study of infant perception by showing the presence of skills much earlier than had previously been imagined.

[2] It is also possible that the socioeconomic differences found in these studies simply represent the degree of familiarity with the stimulus material or testing situation. Neisser (personal communication, April 1990) suggests that nutritional factors, including prenatal ones, may also play a role.

presented with two messages simultaneously through a set of headphones, they can follow the meaning of one or the other but not both (e.g., Cherry, 1953). If they are later asked about the information that was presented to the unattended ear, they can usually recall some of the more general features, such as whether the message was in their own language or whether the speaker was male or female. The fact that they can recall anything at all suggests that the information is being monitored and processed to some extent, but the low level of recall suggests that, unless we actively pay attention, our ability to perceive and process information is extremely limited. As we gain experience, tasks require less attention, and it becomes possible to carry out several activities, such as carry on a conversation and drive a car, at the same time.

As children mature, their attention is directed more and more by acquired schemata, intrinsic interest, and day-to-day fluctuations in motivation and moods. People notice what they want or what they need to see. A person who is hungry, for example, will notice more cues and stimuli associated with food. A person who is sexually deprived will notice more sexual stimuli and so on. Because interests vary with experience, people are tuned to different aspects of their environment. Students who are "turned on" by a course will take in as much as possible, whereas those who are not may find that their minds wander and they recall very little of what is actually said.

People also notice what they have *learned* to see. As people gain experience, their cognitive schemata become more complex and perception occurs more easily. Without the appropriate background knowledge, many of the important aspects of a complex situation may not be noticed at all. Training in a special area makes people more sensitive to certain types of information. A chess master, for example, may notice patterns and configurations that are imperceptible to a novice. A connoisseur can detect ingredients within a sauce or the numerous different dimensions of a fine wine. Perception improves with experience but experience varies from person to person and from culture to culture.

SUSTAINED ATTENTION

People differ not only in their ability to focus their attention and perceive selective features of the environment but in their ability to maintain concentration over time. Some of these differences have already been described. Those students identified as "mastery oriented" by Dweck (1986) are more able to concentrate fully and are less likely to have task-irrelevant thoughts than those described as "helpless." People with an internal locus of control are less likely to let their attention wander, have fewer intrusive thoughts, and are less likely to be distracted by novel, irrelevant information. Training children to attribute their success to ability and failure to lack of effort or strategy also appears to reduce distraction and improve attention.

Other relatively stable personality traits have also been associated with the ability or inability to sustain attention. One of the more important of these is the concept of *test anxiety*. Studies of test anxiety began in the early 1950s (e.g., Mandler & Sarason, 1952; Taylor, 1953), and the concept was quickly incorporated into the research on achievement motivation as a measure of people's "fear of failure." The construct has undergone considerable modification, however, and it can now be explained more in terms of an inability to pay attention.

Earlier models explained test anxiety primarily in terms of psychological arousal. Test anxiety was thought to operate according to the *Yerkes–Dodson principle* (Yerkes & Dodson, 1908), which states that performance is optimal with moderate levels of arousal, and becomes poorer when arousal is either very high or too low. It was assumed that test anxiety increases the level of arousal and that this enhances performance on simple or well-learned tasks and interferes with performance when tasks are difficult. At low levels of arousal, selective attention is poor, and both relevant and irrelevant cues are noticed. At high levels of arousal, a person becomes overstimulated, and well-learned competing responses begin to interfere with performance.

Liebert and Morris (1967) later made a distinction between cognitive and emotional components of test anxiety. The cognitive component, which they described as "worry," is based on a concern about performance and a fear of failure. The emotional component, which had been noted previously, is based on autonomic arousal. Later research showed that the negative relationship between performance and test anxiety was due almost exclusively to the cognitive component in which people express self-doubts and concern about performance and that there is little evidence that arousal interferes with performance.[3]

Liebert and Morris (1967) constructed self-report measures for both components, and subsequent research using these measures identified a number of differences. Worry scores are relatively consistent across time, but arousal tends to peak before the test situation and declines rapidly. Worry scores are reduced by performance feedback, but arousal scores are not. Finally, worry scores are negatively related to performance expectations and actual performance, but arousal bears no consistent relationship with either expectancy or performance. In a review, Wine (1982) points out that the combined results of this research suggest arousal has a transient,

[3] The failure to find consistent differences based on arousal may be due to the nature of the tasks. As mentioned previously, arousal *facilitates* performance when tasks are easy. The inhibiting effects of high arousal can be seen in cases where students prepare for an exam but cannot remember the material during the exam itself. Not only is material lost but relevant and irrelevant information is frequently confused. Once the exam is over, the material comes rushing back. The only remedy for these experiences is to overlearn the material initially, in which case, the arousal could be expected to facilitate performance.

fleeting quality which is confined to the test situation. Worry, however, is a more stable characteristic that interferes directly with cognitive performance, triggers arousal, and plays a major role in maintaining test anxiety.

Wine (1971) drew from Liebert and Morris's (1967) distinction between worry and arousal and developed a *direction of attention hypothesis.* She suggests that those high in test anxiety tend to perform poorly because they divide their attention between self-relevant and task-relevant thoughts. People with high test anxiety express more concern about their performance and are more preoccupied with themselves. Because of this, they switch their attention back and forth and find it more difficult to concentrate. People who are overly concerned about their performance perform more poorly than those with low test anxiety, and these differences are exacerbated if the task is difficult or if it is completed under conditions that increase evaluation apprehension.

The same principle has been applied to other specific forms of anxiety, such as social anxiety, speech anxiety, or dating anxiety (see Wine, 1982). It also appears to be at the root of what Masters and Johnson (1970) call "performance anxiety," which is a sexual dysfunction based on an overconcern with evaluation. There is little evidence, however, that high anxiety in one area is necessarily associated with high anxiety in others. Wine (1982) suggests that test anxiety, for example, may be due to a single traumatic experience with a particular teacher or a particularly stressful examination. She points out, however, that, in some cases, child-rearing practices may establish a general concern with self-evaluation that cuts across specific tasks and interferes with performance in a number of different areas.

There is also some evidence that test anxiety can be reduced by forms of therapy which direct attention away from the self and help people focus on the task at hand (see Wine, 1982). Procedures such as systematic desensitization, on the other hand, that attempt to reduce the level of emotional arousal have consistently produced lower levels of self-reported anxiety but have had little or no effect on actual performance. Some studies suggest that retraining attention can be enhanced if used in conjunction with relaxation training, but the bulk of the research shows that the greatest improvements occur from therapies that focus directly on attention itself.

There is also some evidence that the ability to sustain attention is affected by *television.* The nature and extent of these effects are quite controversial. Although some sweeping claims about the negative effects of television on cognitive development have been made, many of these have been based on idle speculation rather than systematic research. In a thoughtful review, Anderson and Collins (1988) conclude that there is little evidence to support the claims that television has a mesmerizing effect on children, that it causes overstimulation, or that it displaces other more valuable cognitive activities, such as reading or homework. One possible exception is the relationship between television viewing and attention span. This in turn may be partially responsible for the relationships commonly found between television viewing and reading ability, creativity, and success in school.

Singer and Singer (1983; Singer, Singer, & Rapaczynski, 1984), on the other hand, report a series of carefully controlled studies that used TV-logs, direct observation, and parental interviews to study the effects of television on children 4 to 9 years of age. They found that heavy viewing, particularly of violent programs, was associated with poor reading comprehension, less effective use of language, poor school adjustment, restlessness, and aggression. They point out that television uses a variety of gimmicks to capture and maintain attention. These include rapid pacing, frequent interruptions, scene shifts, loud music, and special effects. Foreigners who are unfamiliar with American television, often find these effects disturbing, but American children have come to expect them.

Salomon (1981, 1984) has found that older elementary school children do not find television intellectually demanding. The information processing that occurs as people watch television is more shallow and less elaborate because stories are often quite simple and follow a format that quickly becomes routine. He points to Langer's (1984) research suggesting that people process information quite "mindlessly" when it is familiar, overlearned, or fits into well-developed anticipatory schemata. His research suggests that, although children prefer television, they learn more from reading.

Children do not sit mesmerized in front of their television sets for the most part. The average preschooler looks away from the screen more than 150 times each hour (Anderson, Alwitt, Lorch, & Levin, 1979), and schoolchildren spend about a third of their time engaged in other concurrent activities, such as eating, playing, reading, or doing homework (Anderson, Lorch, Field, Collins, & Nathan, 1986). If they maintain their attention for more than 15 seconds, however, they often show what Anderson, Alwitt, Lorch, and Levin (1979) describe as "attention inertia," which is a relaxed, almost hypnoticlike trance, in which their bodies relax, their heads slouch, and their mouths fall open. Attention inertia is a sign of intense involvement, that may last as long as 10 or 15 minutes.

Research on the relationship between television viewing and attention have typically used correlational procedures in which it is difficult to tease out cause and effect (see Anderson & Collins, 1988). C. Anderson and McGuire (1978), for example, found a positive correlation between teachers' ratings of "impulsivity" and viewing violent programs among elementary schoolchildren. Singer, Singer, and Rapaczynski (1984) used diary recordings of television viewing over a 2-year period. They found that 9-year-old children who viewed a great deal of action-oriented programs tended to be more restless and were less able to sit still for lengthy periods of time.

Both studies suggest that the negative effects of television viewing on attention may depend on the *nature of the programs* and several other studies tend to support this view. Friedrich and Stein (1973) found an increase in tolerance for delay among children who watched the show *Mister Rogers' Neighborhood* but a decrease among those who watched the action-packed series *Batman.* Gadberry (1980) restricted the television viewing of 6-year-old children from middle-class homes and administered the Matching Figure Test of impulsivity 6 weeks later. She found that boys became less impulsive but also noted that restricted children tended to reduce their viewing of violent programs. Those who watched public television programs, such as *Sesame Street* and *The Electric Company,* tended to be less impulsive.

Singer and Singer (1983) have also found that deliberately slow-paced, repetitive preschool programs such as *Mr. Rogers' Neighborhood* or the Australian program *Here's Humphrey* are not only well-received by children but yield gains in cognitive development. The commercial constraints of American television, however, make networks somewhat reluctant to substitute slow-paced programs for more violent adventure series. When the Federal Trade Commission proposed a ban on advertising during children's programs, Congress responded to network pressures by withholding funds that nearly put the FTC out of commission.

There is also some evidence that television viewing influences *reading ability.* Numerous studies have found a negative correlation between amount of television watched and reading ability, but it is unclear whether poor readers simply watch more television or watching television lowers reading ability. Corteen and Williams (1986) examined reading skills in three Canadian towns and found that children without access to television scored higher on tests of reading ability. These differences disappeared after television was introduced. Harrison and Williams (1986), on the other hand, studied the same children and found that those with high vocabulary scores before television arrived tended to watch less television after it became available. They also asked children to list different uses for five common objects and found that the introduction of television reduced the number of creative responses. There thus appears to be a complex interaction between television and reading ability. Those with high reading ability watch less television, and those who watch more television have less reading ability.

Anderson and Collins (1988) suggest that the effects of television on both reading ability and creativity may be mediated by the ability to pay attention. Those who do not pay attention should have more difficulty learning to read and would tend to be less persistent and give fewer creative responses. They also suggest that television may help create a number of habitual responses that are counterproductive when it comes to school. Children who watch a great deal of television are frequently engaged in other activities and may find it difficult to focus when tasks are intellectually demanding. They learn to tune out visual cues and rely on auditory cues, such as dramatic music, that signal what is important and unimportant. They come to associate audio information with visual displays and may find it difficult to listen to audio material without a visual center of focus. And they may come to require more humor, action, and the classroom equivalent of special effects to capture and maintain their attention.

These consequences ultimately affect the way children are taught. Practical guides on how to teach often describe the virtues of multimedia presentations. Teachers and professors are told to intersperse their lectures with slides and videotapes. Such devices are seen as necessary to attract and maintain the attention of students who have grown accustomed to the conventions of TV. Teaching itself becomes a multimedia performance, and more and more time is devoted to teaching teachers how to teach instead of focusing on the content itself. Although such practices have become commonplace, it is not altogether clear whether catering to such deficiencies actually improves performance. It may be that, as lectures become more and more like television, students begin to see them as forms of entertainment and take them less seriously.

The effects of commercial television on people's ability to concentrate help demonstrate that there is no linear relationship between the level of technology and mental ability. Technological advances often force people to develop new skills and change their way of thinking. But technology also simplifies some types of task and makes previous skills unnecessary. Entertainment forms as commonplace as television reach people in large numbers and may ultimately help determine how they perceive and process information. The effects of television need not be all bad. Television can be used to educate as well as entertain. Rapaczynski, Singer, and Singer (1982) have developed a widely adopted program designed to teach children to view television more critically. But the tendency to use gimmicks to capture attention and reduce content to the lowest common denominator may have produced a number of unwanted side effects that few people ever envisioned.

COGNITIVE STYLES

There is also evidence that people and even cultures vary in the way they perceive and organize information. These variations are often described in terms of *cognitive styles* and include traits such as field dependence and independence, reflective versus impulsive approaches to problem solving, and, to a lesser extent, cognitive complexity and the authoritarian personality. Differences in cognitive styles are not limited to perception. They also affect other processes, such as problem solving, memory, and even broader aspects of social interaction. Each of them has some effect on perception, however, so they will be described in this chapter. Cognitive styles are relatively stable tendencies to organize and deal with information in a particular way.

Although research on cognitive styles has focused on a number of different constructs, the area that has probably received the most attention is *field dependence* and *field independence* (see Witkin & Goodenough, 1981, for a comprehensive review). This research began as an attempt to study perception of the upright (e.g., Asch & Witkin, 1948a, 1948b). These studies revealed, quite unexpectedly, that subjects differed markedly in their ability to perceive objects as upright when there were competing visual and body cues and that these differences were systematic and cut across different experimental tasks.

Several measures were developed to compare individuals by separating bodily cues and vision. In one, the *Rod-and-Frame Test,* subjects are shown a tilted luminous square frame in a dark room and are required to adjust a luminous rod until it is pointing up and down. The ability to do this depends on ignoring the titled frame and relying on gravity as the guide. In a second procedure, the *Body-Adjustment Test,* the room itself is tilted and subjects are placed in a tilted chair and are asked to bring their own bodies into an upright position. People who are field dependent have difficulty ignoring the visual cues, whereas those who are field independent do not.

Later tests include the *Embedded Figures Test,* in which subjects are shown simple geometric forms, such as a triangle, and are asked to find them in more complex designs. The ability to find embedded figures requires that subjects break up the larger visual pattern so as to expose the more simple figure contained within. Those who have difficulty finding figures in complex designs also have difficulty adjusting the rod or their bodies when there are competing visual cues and there is a fair amount of consistency across all three tasks.

People who are field dependent have difficulty solving problems in which the solution depends on taking elements out of context or restructuring the visual field. Witkin and Goodenough (1981) claim that underlying this deficiency is a tendency to treat the field in an active or a passive way. Those who are field dependent tend to leave the field "as is," whereas those who are field independent are able to break it up and impose an alternative structure. People who are field independent are more able to perceive items as discrete and to see a pattern in a field with little inherent structure.

Although field dependent and independent people probably do not differ in their perception of most day-to-day events, they do differ in terms of their perception of more complex stimuli. People who are field dependent are dominated by salient features in concept-formation tasks, whereas those who are field independent sample more broadly from nonsalient features and are more likely to solve such tasks when the critical features are not obvious. People who are field independent are less likely to be functionally fixated and are more able to see new problems from a new and different perspective. They respond faster on tests designed to measure the speed of closure and are better able to detect ambiguous figures when there is limited visual information. People who are field independent do better on conservation tasks, such as those used by Piaget, which require people to respond to several different dimensions at the same time. And, not surprisingly, they are less susceptible to visual illusions of self-movement induced by changes within the visual field. There is also some evidence that people who are field dependent are more threatened by stressful or anxiety-provoking material and are more likely to use repression in an attempt to limit the stress (Goodenough, 1976).

There is also some evidence that people who are field dependent are more likely to use a passive "spectator" approach, whereas those who are field independent adopt a more active *hypothesis-testing approach* (Goodenough, 1976). Such an approach implies an active role in generating and testing tentative theories about perceptual differences. When directed to adopt an hypothesis-testing approach, people who are field dependent focus on salient features and attributes rather than sample widely from the range of available features.

People who are field dependent tend to be dependent in other ways. They are less autonomous in their social relations and are more likely to use others as a source of social comparison. College students who are field dependent have been found to be less comfortable in unstructured situations and are rated as less productive by observers (Austrian, 1976). Patients who are field dependent rely heavily on psychotherapists to provide advice and guidance, and psychotherapists appear to respond to individual differences in field dependence by using more open-ended questions for people who are field independent and more questions requiring "yes" or "no" for those who are field dependent (see Witkin & Goodenough, 1981).

Although many studies have found superior performance in learning and memory tasks among those who are field independent, many others have not. People who are field independent learn better under some conditions; those who are field dependent, under others. Goodenough (1976) has pointed out that because people who are field dependent focus on salient features, they should learn more quickly when the salient features are correct—as they typically are. Because of this, it may be more appropriate to see field dependence and independence as a difference in *how* people process information rather than as a difference in *how much* information is acquired.

The superior performance of field-independent people on cognitive and perceptual tasks is matched to a certain extent by their *social incompetence.* Witkin and Goodenough (1981) claim that people who are field independent are more demanding, rude, inconsiderate, manipulating, cold, and distant in their social relationships. People who are field dependent, on the other hand, tend to have a more interpersonal orientation and prefer jobs that require working with other people. They are also more open, warm, affectionate, tactful, accommodating, nonevaluative, and accepting of other people. People who are field dependent are also more accurate in person perception, less likely to show aggression, and more effective in tasks that require conflict resolution. The autonomy or dependence that people show on perceptual tasks occurs in social situations as well and provide a relative advantage to those who are field dependent.

People's overall level of field dependence may depend on early *child-rearing practices.* Parents who encourage independence and autonomy tend to have children with a more differentiated field-independent cognitive style, whereas those who encourage dependence and reliance on parental authority tend to promote field dependence. Parents of field-dependent children are also more likely to use severe forms of punishment to control their children. In families with field-dependent children, parents dominate family interaction. In those with children who are field independent, power relationships are less strongly structured and vary from situation to situation. Mothers of field-independent children have also been found to provide more differentiated feedback by responding differently to different forms of stress, whereas those with field-dependent children respond more quickly but are more likely to use a more general comforting style. These differences occur quite early and are relatively stable over time. Differences in field dependence have been found in preschool children, and although there is a general increase in field independence until the midteens, the relative degree of field dependence remains about the same. Witkin, Goodenough, and Karp (1967) found that scores obtained at age 10 correlated strongly with those obtained at 14, 17, and 24. There may also be *gender differences* in field dependence. Males tend to be more field independent than females (cf. Goldstein & Blackman, 1978). In intact families, girls who identify with their fathers or who see their mothers as more rejecting are also more field independent. Sex differences are more pronounced in cultures where sex roles are stressed.

Finally, there are substantial differences in field dependence based on *culture.* In general, societies that stress conformity tend to promote field dependence, whereas those that stress autonomy and independence foster a more field-independent cognitive style. Societies can be placed along a continuum that runs from "tight" to "loose" (Pelto, 1968). Those at the tight end have more elaborate forms of social control, more role diversity, and more pressure to conform. Those at the loose end have less elaborate social structures and fewer fixed roles, and are more likely to encourage people to "go their own way."

Among the subsistence-level societies reviewed by Berry (1976; Witkin & Berry, 1975), hunters and nomadic societies tend to be more field independent than agricultural communities. Such societies are more loosely structured, place less emphasis on compliance and obedience, and have fewer sharply differentiated sex roles. Berry (1976) combined the various demographic variables associated with tightness and looseness into a single dimension, which typified a nomadic hunting and gathering pattern at one extreme and a sedentary-agricultural type at the other, and found a strong relationship between these patterns and field dependence. Differences were particularly pronounced among females because gender differences were greatly reduced in nonagricultural communities.

A second cognitive style that has received some attention is Kagan's (1971) distinction between *reflectivity* and *impulsivity.* Kagan feels that reflectivity and impulsivity are part of a more general dimension called "cognitive tempo." Differences in cognitive tempo occur quite early. Babies with a slower tempo remain still and look at new objects with fixed attention, whereas those with a faster tempo thrash around, are more easily excited, and look away after a brief examination. These differences also appear to be quite stable. Measures of tempo taken from children at 8 months of age correlate with measures of impulsivity and reflexivity at age 10 (Kagan, Lapidus, & Moore, 1978).

Individual differences in reflectivity and impulsivity are measured by using the *Matching Familiar Figures Test,* in which children are shown a series of 12 pictures and are asked to find an identical match from among 6 alternatives (Kagan, Roseman, Day, Albert, & Philips, 1964). Five of the figures in each set vary in a single feature and only one is identical. Children with a reflective style tend to inspect each of the alternatives and compare them with the original. They tend to examine problems carefully and consider all the alternatives before making a response. Impulsive individuals scan alternatives quickly and select the first one that comes to mind. They respond more quickly but are more likely to be wrong.

Although the main measure of reflectivity and impulsivity is a perceptual matching task, implusivity-reflectivity appears to be a relatively general cognitive dimension that affect a wide range of tasks. Impulsive children make more errors when reading out loud (Kagan, 1965) and perform less well on Piagetian tests that measure conservation of quantity and number (Barstis & Ford, 1977). And a negative correlation tends to occur between impulsivity and standard IQ tests, even though some of the subtests are timed and people are penalized for taking too long. There also appears to be a negative correlation between impulsivity and school performance, even when IQ is held constant (see Messer, 1976, for a review).

Although some of the differences between reflective and impulsive people may be due to differences in the level of physiological arousal, there appear to be social sources as well. Lower-class children tend to be more impulsive than middle-class children. There is also a moderate relationship between reflectivity and field independence. People who are field dependent tend to be more impulsive. It is not clear, however, whether differences in cognitive tempo help promote a more field dependent or independent approach or whether differences in field dependence help foster a reflective or impulsive cognitive style. It is also possible that both stem from similar aspects of child rearing or social circumstances.

Baron, Badgio, and Gaskins (1986) have recently provided a slightly different interpretation of reflectivity and impulsiveness. They point out that although there is a general tendency for errors to be positively correlated with speed (r's around .50 were reported by Messer, 1976), people tend to fall into four general groups. Those who respond quickly and make many errors are considered impulsive and those who respond more slowly and accurately are considered reflective, but there are also people who respond quickly and accurately and those who respond slowly but still make many mistakes. Although impulsiveness is a bad quality almost by definition, neither the rate of response nor the accuracy is sufficient to define it. The hidden assumption is that impulsive people tend to respond *too* quickly and make careless errors that could be avoided if they took more time. These researchers redefine impulsivity as a tendency to stop performing a task before an optimal stopping point has been reached—a tendency to stop prematurely when more time would be helpful in reaching a solution.

Impulsivity can be seen as a self-imposed limitation on information processing. More time could be taken, but it is not. Those who stop prematurely are also more likely to give up when a solution cannot be reached quickly. Baron, Badgio, and Gaskins (1986) suggest that both the tendency to stop prematurely and the tendency to give up quickly have a devastating effect on school performance. Early mistakes are particularly harmful because children fail to develop the basic skills, and as a result, they fall further and further behind. Schools may actually help foster an impulsive approach because tests are often timed, "slow thinkers" are penalized, and the time available for homework is often limited. Baron and his associates present some evidence that impulsivity can be modified within a school setting and that such modification carries over to other tasks not used during the actual training.

A third measure is used to differentiate people is the dimension known as *cognitive complexity.* Cognitive complexity refers to the ability to tolerate ambiguity and think in terms of a number of different dimensions at the same time. People who are cognitively complex, for example, can recognize that their friends possess both good and bad characteristics, whereas those who are cognitively simple have difficulty reconciling inconsistent information (Campbell, 1960). Cognitively simple individuals are more likely to make extreme shifts in their evaluation of other people when exposed to positive and then negative descriptions (Crockett, 1965). People who are cognitively complex are more able to think flexibly, anticipate future events, or alter a course of action when something unexpected happens.

Research on cognitive complexity has focused mainly on adults, and the measures used to determine it, such as George A. Kelly's (1955) Rep tests, are really not appropriate for children. For this reason, there has been very little developmental research on the social origins of cognitive complexity (see Goldstein & Blackman, 1978). Although it may be an important part of information processing, it does not seem to be related to standard measures of intelligence in any systematic way (Bieri, 1955) and there is some suggestion that it may be domain specific (e.g., Ceci, 1990).

Cognitive complexity is related to a broader pattern of behavior associated with the *authoritarian personality* (Adorno, Frenkel-Brunswik, Levinson, & Sanford, 1950). According to these researchers, an authoritarian personality stems from strict discipline during early childhood by parents who make love conditional on behaving properly, emphasize obedience, stress status differences, and are contemptuous of people with less status. These families tend to be father dominated and to have clearly defined and stereotypic sex roles. As a result of having to submit to harsh and arbitrary discipline, authoritarian children develop hostile feelings toward the parents, and toward authority in general, which cannot be expressed openly and are, therefore, redirected onto visible minorities and lower status groups.

Although child rearing shapes authoritarianism to a large extent, it also is influenced by other factors later in life. Leavitt, Hax, and Roche (1953) have shown that teachers help determine children's level of authoritarianism. There are also regional (Greenberg & Fare, 1959) and socioeconomic differences (Brown, 1965). Blacks tend to be more authoritarian than whites (Greenberg & Fare, 1959). The level of authoritarianism also depends on the level of education, with college seniors being lower than college freshman (Plant, 1965; Webster, 1956).

Those with an authoritarian personality tend to be very conventional in their thinking. A fear of impulses leads to a rigid personality based on conventional sexual stereotypes and avoidance of introspection. Lack of psychological insight makes authoritarian people less sensitive to problems within themselves and in other people. Personal relationships tend to be relatively superficial and are perceived in terms of power and status, with an exaggerated emphasis on "toughness" and "strength." Those with an authoritarian personality tend to be domineering toward those perceived as weak but ingratiating and submissive toward those in authority.

The authoritarian personality involves not only a general negative attitude toward the weak and helpless but a way of thinking as well. Although there is little difference between high and low scorers on conventional tests of intelligence, there are considerable differences in cognitive style. Research on the cognitive consequences of authoritarianism has tended to focus on two aspects—rigidity and intolerance for ambiguity (see Goldstein & Blackman, 1978). *Rigidity* is the extent that people are willing to modify their constructs in the face of counterevidence. Rigid people tend to view new situations in terms of previous schemata. They fail to profit from experience because they hold tight to stereotypes and preconceptions. *Intolerance for ambiguity* can perhaps best be defined as the unwarranted imposition of structure on an unstructured situation. It is assumed that authoritarian children learn to reduce conflict and anxiety by keeping some experiences out of awareness. Over time, these habitual responses transfer from social to nonsocial situations, placing a general restraint on perception.

Authoritarians also tend to be more superstitious, read less, and have less intellectual curiosity. Even when they have above-average intelligence, they tend to be relatively rigid in their social thinking and work better with things than people. Their tendency to think in black-and-white terms makes them less sensitive to subtle aspects of the environment and less able to break out of conventional molds when dealing with complex problems.

Again, it should be stressed that differences in cognitive styles are not limited to perception but affect all aspects of information processing. Perceptual differences are important because perception is often the first stage of information processing, and it provides the raw material for other processes which occur later. Variations in cognitive styles help determine the kind of information available, and this capacity affects not just the immediate perception of particular events but people's ability to learn from experience, think and process information, and retrieve information from memory.

CHAPTER 9

Memory

This chapter focuses on memory "skills"—that is, the knowledge and strategies that people develop to make memory more and more effective—but some discussion of memory theories seems necessary to put these skills into perspective. The first is the well-known information-processing model that focuses on the encoding of information and its storage in short-term and long-term memory. The second treats memory as a process of reconstruction in which events that have been previously experienced are recalled. This is followed by an overview of the "memory strategies" used during encoding and retrieval and then sections on how knowledge and higher order, metacognitive processes influence memory. The chapter ends with a general discussion of some of the cross-cultural differences in memory.

There is a close connection between perception and memory. The young infant's ability to recognize objects as novel or familiar depends on memory. The ability to organize perception is aided by the activation of previous schemata, prototypes, and categories stored in memory. Similarly, perception is often the first step in the information-processing sequence that makes memory possible. The ability to encode information adequately determines whether an event will be processed at all or whether it will be processed in a form that can later be retrieved.

Many people equate memory with the ability to recall previous events or pieces of factual information, but there are numerous forms of memory. The concept of memory covers a whole series of complex interrelated systems that serve different functions and behave in very different ways. The ability to recognize objects, for example, depends on memory, and it may occur even when the same objects cannot be recalled. The ability to learn, or to relearn material after it has been forgotten, depends on memory. If a person learns a second language, for example, and does not speak it frequently, about 60% of what was learned is forgotten within the first 3 years (Bahrick, 1984) but it can be relearned more quickly the second time.

Memory is also responsible for everyday "know-how." There is a distinction between declarative and procedural knowledge. Declarative knowledge is based on the storage of factual information that can be verbally described, whereas procedural knowledge depends on knowing *how* to do something. The ability to ski or ride a bike, for example, depends on procedural knowledge. Declarative knowledge can be communicated, whereas procedural knowledge cannot. It may be possible to know how to ski, for example, without being able to describe how it is done. If a person learns to describe this ability, which happens to a certain extent among ski instructors, it then becomes a form of declarative knowledge.

Procedural knowledge seems to involve different circuits within the brain. People who have damaged their hippocampus or amygdala often suffer amnesia and are unable to learn new facts or acquire new information after their injury. They can nevertheless learn new tasks or motor skills. If given a task, such as reading mirror-image writing, making a jigsaw puzzle or even solving a complicated block-stacking brainteaser, their performance improves with practice, even though they have no recall of having seen the task before (Squire, 1987).

MEMORY MODELS

The best known model of memory is probably the *multistore model of memory* proposed by Atkinson and Shiffrin (1968), among others. This model makes a distinction between structural systems, which are discrete storage spaces, and control processes or operations, which process information and move it from one space to another. The storage space consists of a sensory register and a short-term and long-term memory. The sensory register and short-term memory can be thought of as discrete bins that temporarily retain material and through which information passes on its way to long-term memory. Each of the three storage systems varies in terms of (1) trace duration, which determines how long information is retained before it decays, (2) capacity, which determines how much information can be stored, and (3) code or modality, which is the way information is processed and stored.

Information from the senses is briefly retained in the *sensory register* in its original form. Although there is probably a sensory register for each of the five senses, research has focused on vision and hearing. For vision, there is a fleeting photographic memory, called *iconic storage,* that registers an exact copy of the visual scene but lasts for less than a second. The capacity of the sensory registry is apparently quite large. Sperling (1960), for example, showed people three rows of three letters for 1/20th of a second and found that they could recall about half. But when he showed them the same letters and then sounded a high, medium, or low tone that cued them to report only the letters in the top, middle, or bottom row, they had almost perfect recall. If the tone was delayed for as long as a second, however, the iconic image was gone and their recall was greatly reduced. Sheingold

(1973) used a similar procedure with seven geometric forms as stimuli and tested adults and children who were 5, 8, and 11 years of age. There were no developmental differences. The 5-year-olds performed as well as older children or adults. Apparently, iconic storage is well established by age 5 and does not change or improve with age.

There is a similar system for sound, called *echoic storage,* which lasts slightly longer. Sounds, such as speech, seem to linger for three or four seconds before they disappear. This allows people to retain information conveyed by speech so that it can be processed as a unit. It also explains why people sometimes misunderstand a message initially but grasp the meaning while they are asking the person to repeat it.

One of the central assumptions of the multistore model of memory is that information "moves" through memory in a number of discrete steps. Information from the sensory register may be passed on to *short-term memory,* where it is retained for 10 to 30 seconds. Information in short-term memory occurs primarily in verbal form because older children and adults have a tendency to talk with themselves as they process information.

The *capacity* of short-term memory is limited to about seven items, plus or minus two (Miller, 1956). This may be seven numbers, seven letters, seven words, or seven larger chunks of information. There are a number of qualifications, however. Short-term memory is better for random numbers than random letters and better for spoken than visual information. The capacity of short-term memory is also determined, at least in part, by the rate of articulation. As mentioned previously, Welsh children have a lower digit span than English speakers because it takes longer to pronounce numbers in Welsh (Ellis & Hennelley, 1980).

Short-term memory is often referred to as immediate memory or *working memory* because this is where information is actively processed and transformed. There are a number of control operations, such as chunking, rehearsal, organization, elaboration, and the use of mental images, which can be used to make the material more meaningful and help retrieval. Developmental differences in the ability to process information may be due to differences in the capacity of short-term memory, differences in the control operations, differences in the amount of background knowledge available to make the material more meaningful, or some combination of these.

If the information in short-term memory is acted on and made meaningful, it passes into *long-term memory,* where it is permanently stored. Information in short-term memory is stored acoustically (by sound), but information in long-term memory is often stored semantically by meaning (Baddeley, 1982). Long-term memory has an unlimited capacity, but it may not be possible to recall items stored in long-term memory. The extent of recall often depends on the meaningfulness of the original material, the number of associations or retrieval cues, and whether or not there is interference from similar material learned formerly or later. The disruptive effect of previous material on the recall of new information is called "proactive interference," whereas "retroactive interference" occurs when new information makes it difficult to recall things that have been learned previously.

If we put aside the sensory register (whose function in everyday memory is not altogether clear), the distinction between short-term and long-term memory merely refers to the distinction between the contents of immediate experience and what is commonly thought of as memory proper. The processes occurring during short-term memory are by no means confined to memory. They play a major role in perception, language comprehension, thought, and various forms of problem solving, and correspond to what William James (1890) referred to as the "specious present"—that brief period of time in which items of information are processed together. It would be wrong, however, to assume that information processing stops once information is moved into long-term memory. There is often a quite slow, mostly unconscious process of consolidation that may take several years—people who suffer retrograde amnesia often experience an inability to process new information and a memory loss for events occurring one or two years previously. This slow process of consolidation appears to occur in the brain region known as the "hippocampus," which functions somewhat like an off-line computer (see Squire, 1987). Once information is learned and consolidated, the emphasis switches from encoding to retrieval.

While the multistore memory model focuses primarily on the encoding and storage of information, several quite famous theories focus on *retrieval* (e.g., Bartlett, 1932; Piaget & Inhelder, 1973). Once of the most active researchers in this area is Ulric Neisser (1967, 1982, 1984, 1986a, 1988). For Neisser, remembering is a process of *reconstruction* that uses some of the original material as a source of raw data. He rejects what he calls the "reappearance hypothesis," which assumes that the current memory is a reproduction of the original event. Reconstruction rather than literal recall appears to be the rule and this can be seen quite clearly in eyewitness testimonies where leading questions can cause witnesses to substantially alter their recollection of the original experience. Remembering is more like an act of problem solving than entering a storage bin, and recall may be enhanced or inhibited by current knowledge.

Neisser has recently extended his view of memory by drawing from the distinction between *episodic* and *semantic memory* and stressing the "nested" or hierarchical quality of experience. Episodic memory is based on the recall of personal experiences and events. These include the numerous day-to-day events that were sufficiently significant to leave a lasting impression. Semantic memory is based on knowledge about the meaning of words and typically does not include the original experience. I may know what the word "memory" means, for example, but I cannot recall when it was first learned. The concept of semantic memory was introduced by Quillian (1968) but broadened by Tulving (1972) to include any kind of stable, nonpersonal knowledge, such as chemical formulas or the temperature of cities in different parts of the world.

Neisser (1988) points out that this type of memory can be extended to larger events, such as knowledge of scripts and more extended episodes. *Scripts* are knowledge of how to behave in typical situations, such as going to the doctor or eating at a restaurant (see Schank & Abelson, 1977). They are based on schemata for events, abstracted through experience with a number of similar situations. As with words, the particular contexts in which the scripts were learned are eventually lost but the general knowledge remains. Knowledge for scripts is based on the underlying structure or common features rather than the specific details. When a number of experiences are similar, the common structure is retained although the details may be lost or confused.

Neisser points out that events that lead to the development of *scripts* are nested within larger events and contain smaller, more specific details. Recalling a particular event is not a matter of reviving the original experience but also tends to be hierarchical, and memory can move up to capture the larger context or down to focus on smaller, more particular details. A side trip to Italy, for example, may have taken place during a summer vacation and it may have involved eating at a particular restaurant or staying in a particular hotel. Eating at the restaurant itself involved a number of more specific features, such as the taste of the food and the behavior of the waiter. Recollection of these events cannot include every aspect of the original experience. Attention is often focused on one level at a time, but even when we cannot recall the larger context or the particular details, we still know that they occurred, and this confidence makes reconstruction an almost inevitable part of any memory.

The nested quality of experience helps explain some of the *inaccuracies* which occur in memory. Mistakes occur because people do not pay attention or confuse one event with a similar experience that happened at a different time. Information that is highly organized and tied into an extensive body of knowledge is much more resistant to forgetting, whereas isolated bits and pieces of information are quite vulnerable and easily confused. When we experience the same type of event over and over again, we may lose all but the most general features and not be able to distinguish one specific instance from another.

To illustrate this, Neisser (1988) draws from his own research, which compared John Dean's testimony during the Watergate hearing to White House tapes of the actual conversations (Neisser, 1981). During his description of one encounter, John Dean was wrong on almost every detail. He put words into people's mouths that were never spoken, and he omitted details that actually occurred. John Dean was basically right about what was happening at the White House, however. He simply could not recall with perfect accuracy who said what to whom.

The hierarchical structure of memory means that memory is subject to certain types of distortions. General memories persist even when the details are lost. Specific events happening on similar occasions are frequently confused, and we are likely to insert common events into situations in which they failed to occur. Scripts not only serve to organize but can also bias our recall. Surprisingly, inaccuracies in recall are often accompanied by visual images of events that never occurred. These help to increase people's confidence in their own memories and help explain why there is little correlation between confidence and accuracy (Loftus, 1979).

MEMORY STRATEGIES

Although the structure of memory seems to be fixed relatively early, there are additional changes in control processes and mental operations which allow greater and greater recall. Improvements in memory are based on three interrelated sets of changes, which may or may not occur as people mature. The first is the development of memory strategies, such as chunking, rehearsal, organization, elaboration, and the use of mental images. The second is the acquisition of knowledge within a particular domain, which makes access to information faster and more orderly. Third, there is the development of higher order cognitive skills, including the person's awareness of the contents and limits of his or her own memory and conscious access to a number of memory strategies that can be deliberately used to process information. These three areas are very much related, but it may be useful to treat them separately before mentioning where they overlap.

Memory strategies are processes that can be placed on a continuum representing the extent of effort and the deliberate manipulation of information (Bjorklund, 1989; Hasher & Zacks, 1979). At one end of the continuum are *automatic processes,* which occur without conscious awareness or deliberate control. They occur so effortlessly that they are not noticed. At the other extreme are *conscious strategies* that can be deliberately used to process information and aid recall. These require some effort and interfere with other processes because attention is limited. The most common strategies occur during encoding and include chunking, rehearsal, organization, elaboration, and the use of mental images. Other strategies can be used during retrieval, whereas still others involve the development of elaborate artificial strategies based primarily on the use of mental images.

The distinction between deliberate and automatic processes is useful, but it may obscure a second distinction that is equally important. Some mental strategies are initially used deliberately and become more and more automatic with practice. These include rehearsal, elaboration, and the deliberate use of mental images. Others, such as chunking, organization, and the spontaneous use of mental images, occur more or less automatically from the start. Improvements are often based on knowledge within a particular domain. What makes these differences more difficult to grasp is that the same strategies may be used either deliberately *or* occur more or less automatically. Chunking and organization can be used as a deliberate strategy to aid recall, but they also occur automatically as people gain experience within a particular domain.

One of the major assumptions of the information-processing approach is that *attention is limited.* The capacity of working memory among adults is limited to between five and nine items, depending on the person and the nature of the material. Case (1985) has suggested that deliberate strategies to enhance memory may further limit this capacity. Case makes a distinction between storage space and operating space. Storage space is the hypothetical amount of space available for storing information, whereas operating space is the space needed to carry out a particular operation. As people adopt additional strategies and carry them out, the processes become more and more automatic and require less and less effort. This leaves more space available for the storage of new material. Young children take more time to process information and, therefore, use more of their limited space. Other factors, such as familiarity, also influence the speed of processing and the amount of space available.

One of the simplest ways of getting over these limitations is to *chunk* information. Chunking groups individual items into larger units so that they can be processed together. Any string of digits such as,

1 9 0 2 5 3 9 7 3 9 2

is hard to remember one digit at a time because it exceeds the limits of working memory. The same items can be more easily recalled if they are grouped into two digit numbers,

19 02 53 97 39 2

and they are easier still if placed in sets of threes or fours.

This is an example of a deliberate strategy used to recall an otherwise bewildering array of random numbers, but chunking can also occur automatically. The preceding numbers are quite easy for me because they are my telephone number, complete with the area code for Nova Scotia and the mandatory 1 for long distance. It has become essentially one long number. As people gain experience in a particular area, they become more able to chunk previously separate items into meaningful groups. Chess masters, for example, can reconstruct the layout of most pieces on a board after a five-second inspection, because they recall the pieces in patterns (Chase & Simon, 1973). If the pieces are randomly arranged, however, the recall of chess masters is no better than that of an average person.

A second strategy is *rehearsal.* Items in working memory can be repeated over and over again until they are easy to recall. Rehearsal helps memory because it keeps items in working memory and it facilitates recall. The rehearsal strategy develops gradually. Flavell, Beach, and Chinsky (1966) studied verbal learning in children and found that only 10% of the 5-year-olds used rehearsal. By age 10, this figure had climbed to 85%. In a similar vein, Wagner (1978) carried out a study of schooled and unschooled people in Morocco and found that rehearsal skills developed gradually between 7 and 19 years of age for those in school but that they were uncommon at any age among those not in school. Children as young as 5, however, can improve their memory through rehearsal training (Asarnow & Meichenbaum, 1979; Ornstein, Naus, & Stone, 1977).

A third, even more effective strategy is *organization.* Separate items of information can be organized either by drawing from the inherent structure of the material or by superimposing an artificial structure onto what would otherwise seem to be a meaningless array of disconnected facts. The names of the Great Lakes, for example, can be remembered by using the acronym, HOMES, which contains the first letter of each of the lakes—*H*uron, *O*ntario, *M*ichigan, *E*rie, and *S*uperior. The colors of the rainbow can be recalled by remembering the name ROY G. BIV (*r*ed, *o*range, *y*ellow, *g*reen, *b*lue, *i*ndigo and *v*iolet). Such techniques use an artificial structure, but they can be effective.

More common is the tendency to organize information on the basis of categories and common features. If adults are shown a list that contains a number of related words, such as dog, cat, horse and so on, they spontaneously associate them with the more general concept of animal and remember them as clusters during free recall. Young children also tend to process words in terms of categories but they do not use these categories to organize information (Brainerd, Kingma, & Howe, 1986). They seem to process words on an item-by-item basis but miss the common link that ties individual items together. The ability to use inherent properties to organize information also improves with age.

Organization of material also improves with *experience.* Ceci (1980), for example, gave children 4, 6, and 9 years of age extensive training that included common characteristics of animals such as diet and habitat. Children with such training showed superior recall. Chi (1978) compared a group of children from a local chess tournament with college-educated adults who could play chess but were not experts. Although the adults could recall more digits than the children, the children were better able to recall the position of chess pieces on the board. The effects of knowledge on information processing will be described in more detail in the next section but it shows that so-called techniques, such as chunking or organization, can be used more or less deliberately but can also occur more or less automatically once people gain a sufficient level of knowledge.

A fourth strategy is *elaboration.* When we elaborate, we also go beyond what is actually given and relate the material to information that has been previously acquired. The most common form of elaboration is probably the tendency to make new material personally relevant by tying it into our own experience. We remember people, places, and events better when they are personally relevant. To a certain extent, this is an inevitable consequence of having to rely on our own experience to understand new material, but it can also be used deliberately to aid recall. Students in introductory psychology courses, for example, are often encouraged to make the material personally relevant by relating it to their own lives. Young children are not very good at relating new information to what they already know, but this too improves with practice.

Mental images provide a fifth strategy. Words that lend themselves to visual images are more easily recalled. It is easier to remember concrete objects than abstract objects because concrete objects can be processed in terms of both images and meaning. The visual memory system appears to be separate from the acoustic system used in working memory (Baddeley, 1982). It draws from information in long-term memory but involves different circuits within the brain. There is a complex interplay between verbal and visual recall. When a person is trying to recall an event, there is a strong tendency to provide a running commentary, thus turning a visual task into a verbal one. This dual coding tends to aid recall. The deliberate use of mental images to recall verbal material, which will be described later, is a powerful mnemonic technique that can be used to substantially improve memory.

These memory strategies should be seen as complex cognitive skills that develop gradually over time. Some, such as chunking and organizing information, become so automatic that they are not even noticed. Others depend on the recognition that recalling certain types of information is difficult and is facilitated by the deliberate use of memory strategies. Children appear to go through a number of distinct stages as they develop these skills. Very young children do not spontaneously use any strategy and seem incapable of using them effectively even when instructed to do so. This inability is known as a *mediational deficiency* (Reese, 1962), and it involves a lack of mental ability, where the required cognitive skills have yet to develop. Slightly older children can profit from instruction, but they do not spontaneously use memory strategies without being prompted. This is known as a *production deficiency* (Flavell, 1970). Children during this stage have the ability to use cognitive strategies, but for some reason, do not.

This seems to be followed by a third stage in which children begin to deliberately use cognitive strategies, but they produce little or no real gain in memory. This has been described as a *production inefficiency* (Flavell, 1970), and it is typically seen as a transition period in which skills are being developed and tried for the first time. These early deliberate efforts require so much attention that they seem to interfere with other processes and use space that would normally be allotted to items in memory. These skills become more and more automatic with practice and eventually are carried out so effortlessly that they place little demand on attention. In fact, after they have become automatic, they can be disrupted by paying too much attention.

Although most of the improvements in memory occur because people learn to process or encode information more effectively, some techniques can be used to aid *recall*. The most common technique is probably a conscious search for items in memory. If someone is trying to remember where they were on a particular day—say 1 July 1984—they can start by trying to remember the year, then the summer, then important events surrounding the particular day. This self-conscious search takes advantage of the fact that information in memory is hierarchically organized, with more specific details embedded within a larger context (Neisser, 1988). The mind can grasp for retrieval cues in a more or less systematic fashion.

One of the reasons that recognition is typically easier than recall is that recognition provides its own retrieval cues. This was shown dramatically in a study carried out by Bahrick, Bahrick, and Wittlinger (1975). They asked people to remember former classmates 25 years later and found that most people had limited recall. When they were shown names and faces, however, the recognition rate was 90%. Young children profit more from retrieval cues than adults, presumably because older children and adults generate their own.

Physical locations can serve as a retrieval cue, and memory can sometimes be improved by returning to the place where the event occurred. This is why witnesses are often taken to the scene of the crime. The physical location can elicit a number of conscious or unconscious associations that may aid recall. This is known as the *context effect*. Godden and Baddeley (1975), for example, gave a list of words to scuba divers who were either underwater or on land. Words that were presented underwater were recalled better underwater, whereas those presented on land were recalled better on land.

Memories also appear to be *state dependent*. When people learn words while feeling sad, they tend to recall those words better when in a sad mood, whereas words learned when people are happy are recalled better when they are happy (Bower, 1981). This is why depression has such a negative and, to a certain extent, self-perpetuating effect. Lewinsohn and Rosenbaum (1987) found that people who are currently depressed recall their parents as more rejecting, punitive, and guilt promoting, whereas those who were formerly depressed do not differ from people without depression. The context effect and state dependence seem to be limited to recall, however. Recognition is not affected. In both cases, the effect seems to be due to the presence of additional retrieval cues that "jar" recognition and aid recall.

Memory for specific items and events can also be improved by *periodic recall*. The best way to remember an event is to recall it often and reorganize it on each subsequent occasion. One of the reasons that our memory for past events seems so spotty is that we are more likely to remember pleasant events and ignore or deliberately avoid the unpleasant ones. This is one reason childhood seems so charming to most people. As adults, our childhood seems to have been periodically punctuated with the pleasant experiences of birthdays and Christmases and the genuine surprises that took on a special significance. What is missing is the memory of the hum-drum and routine events that occurred in between.

Knowledge of memory strategies occurring during encoding and recall helps to explain one of the most baffling of all psychological events—*forgetting*. If we ask, Why do we forget? the most obvious answer is that the information was never processed properly to begin with. The large number of memory strategies shows how difficult it is to get information into long-term memory. Much of what we experience is never processed effectively because it is considered too trivial or insignificant. People also tend to overestimate their ability to remember information and, therefore, fail to make the necessary effort to process it effectively in the first place.

Sigmund Freud was one of the first to note the role that language plays in memories. For Freud, information was conscious when it was accompanied by a verbal description, and it was unconscious when it was not. Unconscious material is deliberately avoided because of its painful associations or threat to a person's sense of self-worth. Freud developed the technique of free association (or the "talking cure") as a way of bringing previously unconscious ideas into awareness. During free association, patients are encouraged to talk about anything and everything that comes to mind. As they become more comfortable with the psychoanalyst and the situation, they become more willing to dig deeper and describe events that have been previously repressed.

Although Freud had a profound knowledge of many things, he had a poor understanding of memory. Freud assumed that virtually all of life's experiences, including those from early childhood, are stored in memory. Those that could not be recalled were kept unconscious through a deliberate process of repression. Repression was seen as a failure to retrieve items from memory because they were too painful or had other negative associations. Freud was misled to a certain extent by the power of his own memory, which seems to have been rather remarkable. For a time, Freud's theory received some support from the electrical stimulation studies of the Canadian brain surgeon, Wilder Penfield (see Penfield, 1975). Penfield claimed that his patients often experienced quite vivid and detailed experiences of previous events while undergoing surgery. Closer scrutiny of this research by Neisser (1967) and, later, Loftus and Loftus (1980), however, showed that such reports were extremely rare, and when they did occur, the contents suggested that they had been invented rather than recalled.

Nor has there been much support for Freud's theory of *childhood amnesia*. Most people have few recollections of events occurring before age 3 or 4. One reason may be that young infants do not have and cannot use the various memory strategies commonly used by adults. Infants also lack well-developed language skills which can be used to provide a running commentary on what is happening and consolidate events. Infantile amnesia is not due to a poor memory, however. Nelson and Ross (1980) have shown that 2-year-olds can remember events for as long as a year, even though they only happened once. Childhood amnesia also appears to be limited to episodic memory. There is little or no loss of memory for words or procedural knowledge. An alternative explanation is that the young child's experience of the world is so fundamentally different from that of adults that it is no longer recognizable. The development of a more mature perspective makes it difficult to go back and reconstruct events that were processed from a child's point of view (Neisser, 1967; Schactel, 1947).

Not all memories require such deliberate attention and processing, however. Some experiences are so surprising, painful, or significant that they are thrust upon the mind almost as an intrusion. If we suffer the loss of a loved one or undergo a particularly humiliating experience, we may be haunted by the memory for some time. Eventually, the intensity of our feelings weakens and the memory occurs less often, but it nevertheless seems to be "etched" permanently on our brain.

A similar process seems to occur during *flashbulb memories.* Most Americans over 40 believe that they can tell you where they were and what they were doing when they heard the news of John F. Kennedy's assassination. Flashbulb memories, while they are vivid and detailed, need not be accurate. Neisser (1982), for example, has described how he recalled hearing a news flash interrupt a baseball game on the radio when the Japanese bombed Pearl Harbor. The problem with this memory is that the bombing of Pearl Harbor occurred on 7 December 1941, long after the end of that year's baseball season. Thompson and Cowan (1986) later pointed out that what Neisser actually heard was the interruption of a football game being played by two teams whose baseball and football names were similar—the New York Giants and the Chicago Bears. Neisser (1986a) later admitted that this was probably what had happened but that the confusion of a baseball and football game does not support the view that flashbulb memories are accurate and indelible imprints of the original experience. Interest in flashbulb memories has recently been rekindled by the space shuttle *Challenger* disaster. Recent analysis of people's memory for this disaster taken one day and several years after the explosion casts serious doubts on the accuracies of such memories (Neisser, 1990). Flashbulb memories may *seem* accurate, but there is little correlation between the accuracy and the confidence of those recalling the experience.

THE KNOWLEDGE BASE

Although memory is partially determined by the development of memory strategies, it is also strongly influenced by what people already *know.* There is a complex interplay between memory strategies and people's base of knowledge. As people gain experience within a particular domain, meaningful chunks of information contain more items and are more tightly organized. People sometimes develop additional memory strategies that are domain specific. Processing information becomes faster and occurs more easily. Some changes are so automatic that they occur without conscious awareness. People do not understand why they are so proficient in some areas. Some things just seem easy.

The relationship between knowledge and memory was obscured for many years because of the tendency to use novel and sometimes nonsensical material, but there has been a recent shift in emphasis. Not only is there more concern about ecological validity, but there is also a tendency to compare people who know a great deal or very little. Differences between experts and novices can be explored under laboratory conditions, and when they are, they tend to be quite striking.

One of the first attempts to systematically explore the role of knowledge on memory involved recall of the position of chess pieces. Chase and Simon (1973) compared the memories of experienced and nonexperienced players by showing them configurations of chess pieces for five seconds. A chess master or grand master could reconstruct the position of 25 pieces with 80% to 90% accuracy, whereas novices recalled the position of only about six pieces. If the pieces were placed on the board in a random manner, the differences disappeared.

Differences between players were based on several factors. The most noticeable were differences in the size of *chunks* used to process information. Although both experts and novices had the same number of chunks in working memory, the chunks of novices were based on individual pieces, whereas chess masters tended to chunk items by configurations, averaging three or four pieces each. These were patterns that had become thoroughly familiar through thousands of previous encounters. A number of individual pieces were now perceived as a unit and recalled in the same way. Simon (1974) points out that the ability of a chess master to remember the positions of 25 pieces after a brief exposure is no more mysterious than a reader's ability to recall a sequence of 25 letters—provided they are arranged in terms of four or five familiar words. It is important to realize that the expert chess players were not using chunking in any deliberate sense. The configurations of chess pieces that they noticed had built up gradually over time and were the result of thousands of previous experiences.

A similar study was carried out by Michelene Chi (1978). Chi compared children from a local chess tournament (mean age = 10.5 years) with research assistants and graduate students with some knowledge of chess. When tested for digit-span memory, adults showed the typical pattern of superior recall. When the two groups were tested on their ability to reconstruct the position of chess pieces after a brief exposure, however, the children were much better. The children also took fewer trials to learn the positions of all the pieces on the board. Chi attributes these differences not only to the pieces being grouped into larger chunks by the more experienced children but also to the overlapping features shared by the chunks themselves. A single piece was often included in several different configurations, so that recalling one served as a retrieval cue for the second one.

Chi (1978) also found that several of her subjects used task-specific memory strategies. Two subjects said they memorized the position of pieces by taking advantage of the symmetry of the game. If they found a rook in one corner, for example, they would look to see if there were rooks in the other corners as well. If there were, then all the rooks were processed as a unit. This unit could then serve as a retrieval cue for other configurations containing one of the rooks.

A rather exceptional example of a domain-specific memory strategy can be seen in the cases of SF and DD, two long-distance runners who learned to extend their digit span from 8 items to 84 and 106, respectively. Both men learned to recall series of digits as running times, ages, or dates, and yet their memory span for other items, such as words and letters, was not affected (see Staszewski, 1990). The amount of knowledge can sometimes influence the use or nonuse of even basic memory strategies. Bjorklund (1988), for example, found that children were more likely to use higher order categories to organize typical words (e.g., *clothing* for shirt, pants, dress) than atypical words (e.g., *clothing* for shoes, ties, socks).

During the 1980s, a number of studies were carried out comparing the memory of experts and novices in various domains. One of the more relevant for our purposes was a study carried out by Carol Walker (1987), who looked at the relative importance of knowledge and general ability. Walker divided soldiers into four groups on the basis of their scores on the Army aptitude test and their knowledge of baseball. Those who knew a great deal about baseball recalled and recognized more information about a hypothetical game and were better at inferring information that was implied but not explicitly stated (such as the score or the number of runners left on base). The level of general ability had absolutely no effect. Low-aptitude soldiers with a knowledge of baseball performed significantly better on all tasks than high-aptitude soldiers who did not know the game well. This seemed somewhat surprising given the common assumption that people's ability to perform well is based on a combination of specific knowledge and general ability. High-aptitude soldiers with little knowledge of baseball often handicapped themselves by using counterproductive strategies, such as memorizing the batters' names.

Experience also helps determine the speed of information processing, which is often assumed to be one of the most basic aspects of "general intelligence." Ceci (1990) has recently challenged the claim that there is a single underlying factor, based on the speed of information processing, that determines general ability. He reviewed a number of studies showing that familiarity plays a major role in even such elementary skills as letter recognition and mental rotation. These skills are sometimes seen as so elementary that they cannot be based on experience and they are, therefore, seen by some as the basic building blocks for higher order skills (e.g., Eysenck, 1982). Those who accept these theories believe that there are basic physiological limits which constrain intelligence, beyond which people cannot pass. A number of studies fail to support this position. First, tests of mental speed are not highly correlated (Ceci, 1990). People who perform rapidly on one task may perform poorly on another. Second, speed of recognition is strongly influenced by practice (e.g., Regian, Shute, & Pellegrino, 1985). As people gain experience, they can recognize and process information more swiftly. Familiarity continues to influence processing speed well into adulthood. Words that are encountered more frequently by adults can be named faster and recognized and recalled more easily.

In short, knowledge tends to affect almost every aspect of memory. As people gain experience within a particular domain, this increase in knowledge improves memory by (1) increasing the speed of recognition and recall, (2) increasing the size of meaningful chunks that can be processed as a unit, (3) providing an organized framework within which information can be processed and retrieved, and (4) promoting the development of deliberate memory strategies. The relationship between knowledge and various cognitive skills is so profound that it calls into question the very concept of general ability. Walker (1987) suggests that what passes for general intelligence is simply a great deal of experience within a number of different domains. Tests of general intelligence sample knowledge and ability by including a variety of different tasks, but they may underestimate people's intelligence when their knowledge is limited to one or two areas. Such tests do not actually measure what people *can* do given sufficient background knowledge and motivation. If someone wanted to develop tests that measure what people are actually capable of doing—in terms of comprehending, organizing, and recalling information—it would be better to use tasks about which people already know a great deal, for it is only on such tasks that people have fully developed their skills.

Walker (1987) goes on to suggest that, because expertise develops gradually and because skills become more and more automatic with practice, people who are highly knowledgeable in only one or two areas may not recognize their own ability or be able to transfer their skills from one domain to another. Understanding baseball, for example, may seem easy, natural, or even mysterious. But it only seems that way because people have acquired both the necessary skills and sufficient background knowledge to make information processing fast and efficient. People with high levels of general ability simply generalize skills across domains, whereas those with less general ability do not. The absence of transfer across domains is not a sign of low intelligence, however, because the person whose expertise is limited to one or two areas may perform exceptionally well within those domains. The person fails to develop general strategies to aid memory because memory strategies within a particular domain have become so automatic that they occur without forethought or awareness. Transferring mental strategies from one domain to another often depends on the development of a higher order, metacognitive awareness of their utility.

METAMEMORY

The answer to Flavell's (1971) famous query about "What develops?" in memory seems to be memory strategies, a knowledge base, and metacognitive memory processes. The latter is often referred to as "metamemory." Although there are numerous variations in the way metamemory is conceptualized and described, the most useful definition seems to be based on a distinction between conscious awareness of our own memory and our knowledge and control of the various memory strategies used to aid information processing and recall. Awareness of our own memory includes a knowledge of what has and has not already been learned. If we feel that some piece of information has been previously learned but cannot be recalled, this may motivate us to systematically search our memories to retrieve it.

Perhaps more important is the knowledge that our own memory is limited and that some types of information must be deliberately processed to be recalled. Young children do poorly on memory tasks because they do not have a good grasp of their own ability. They tend to overestimate their ability to recall information and fail to recognize their own limitations. This can be seen in two similar experiments carried out by Flavell, Friedrichs, and Hoyt (1970) and Yussen and Levy (1975). When preschool children were asked to judge their memory span for the number of items that they could hold in working memory, they estimated that they could recall about eight items but the actual number was only four. The discrepancy between what children thought they could recall and what they actually recalled decreased throughout elementary school, as the estimates became more and more realistic. In a similar vein, Kreutzer, Leonard, and Flavell (1975) asked children questions about their ability to recall information and found that some of those in kindergarten claimed that they never forgot anything.

This suggests that young children fail to use memory strategies, even when they have the ability, because they are unaware that such strategies are necessary. Wellman (1977) claims that young children have little knowledge of their own ability because they do not effectively monitor their own performance. It is also possible that errors frequently go unnoticed because there is little direct feedback about their lack of ability. This changes radically once children enter school. The conscious and deliberate use of memory strategies seems to develop partly in response to the increasing demands placed on memory in school. Other improvements include the knowledge that it is easier to recall the gist of a message than to recall it verbatim, that more time is needed to study material when there is a long delay between learning and recall, and that items can be recalled easier if they are arranged in categories (Bjorklund, 1989).

People's self-conscious awareness of their own memory and its limitations is a form of declarative knowledge, but a second aspect of metamemory involves both declarative and procedural knowledge. This is based on an awareness and deliberate use of the various memory strategies described previously. Some of these, such as clustering and organization, may occur more or less automatically and are a direct result of how much a person knows. Others, such as rehearsal, elaboration, and the manipulation of mental images, can be deliberately used. Additional mental strategies include studying information in the order of importance, allocating more time to material that is not already well learned, surveying and reviewing material, and questioning oneself about what is known.

Metamemory, like memory strategies in general, develops gradually over time. As mentioned previously, young children do not spontaneously use memory strategies because they overestimate the power of their own memory and do not notice cases where they forget information. In Western societies, one of the first skills to develop is rehearsal. Young children learn to repeat material over and over to aid recall. Somewhat later, children begin to use higher order categories to organize material. Still later, simple strategies appear to give way to more complex strategies. Pressley & Levin (1977), for example, found a shift from rehearsal to elaboration somewhere between the fifth and ninth grades. They also found that fifth graders who used elaboration outperformed ninth graders who did not. A similar pattern has been found by Waters (1982) for students in the eighth and tenth grades. Waters also found that tenth graders who used elaboration recalled significantly more items than eighth graders using the same strategies, suggesting that strategy *effectiveness* improves with practice (and/or age). There, thus, appears to be a complex relationship between metaknowledge, strategy effectiveness, and strategy use. Numerous studies have shown that metamemory strategies can be taught, and when they are, they often have a lasting effect and transfer to other, quite different domains (see Pressley, Borkowski, & O'Sullivan, 1985).

Although some reviews have found very little relationship between metamemory and actual performance (e.g., Chi, 1984; Cavanaugh & Perlmutter, 1982), a metaanlysis of 47 separate comparisons carried out by Schneider (1985) showed a strong and quite significant effect. The average correlation between measures of metamemory and academic performance in these studies was .41. Some aspects of metamemory contributed more heavily to the overall correlation than others, but all in all, both aspects of metamemory seemed to be positively related to performance. John Borkowski and his associates have found that, once metaknowledge is available, metamemory predicts memory performance better than more general factors, such as verbal intelligence or cognitive style (Borkowski, Peck, Reid, & Kurtz, 1983; Borkowski, Reid, & Kurtz, 1984).

The lack of correlation in some studies is based on methodological problems, such as the measurement of metamemory or a narrow range of memory tasks, but there are also other (nonmethodological) reasons to expect a limited relationship in some cases. First, as previously mentioned, children often do not use memory strategies because they do not recognize the limits of their own memory. *Some* recognition of these limits would seem necessary to motivate people to develop memory strategies, but a perfect correlation between accuracy and performance is unlikely. Adults do quite well on recognition tasks, although they underestimate their own ability (Levin, Yussen, DeRose, & Pressley, 1977). Research on problem solving, which will be described in more detail in the next chapter, suggests that experts within a particular domain tend to be *less* self-consciously aware of their own mental operations (e.g., Simon & Simon, 1978). A strong, well-integrated knowledge base sometimes reduces or eliminates the need for deliberate strategies. Despite these disclaimers, studies that have used knowledge of memory as their measure of metamemory have typically found a positive relationship between knowledge and performance—those with good memories are more aware of their own memories and its limitations.

Researchers would also not expect a perfect correlation between memory and the deliberate use of memory strategies. As mentioned previously, there may be a phase in which memory strategies begin to occur but are not well developed and interfere with memory. This has been called a *production inefficiency,* and although the evidence is limited, it may represent a hurdle beyond which people must pass to use memory strategies efficiently. The failure to find production inefficiencies may be due to their brief duration. If they occur, they would occur just once during the development of each memory strategy.

A metacognitive awareness of memory strategies also does not guarantee that they will be used to aid recall. The use or nonuse of a memory strategy also depends on the nature of the material, the amount of motivation or some extrinsic goals. The failure to use memory strategies has been described as a *production deficiency,* and it appears to be quite common among children. Adults may also fail to use well-learned memory strategies if they are insufficiently motivated or unsure of the need. Most adults in Western societies, for example, know that rehearsal aids recall, and yet they may fail to use it in a situation as commonplace as remembering a name. At the time of introduction, it seems unnecessary and its usefulness is understood only after the name is forgotten. Adults, like children, frequently overestimate their ability to remember and discover their limits only after it is too late.

CROSS-CULTURAL DIFFERENCES

The study of memory development in Western societies gives the impression that memory strategies and metacognitive skills develop in a strict sequence as children mature. But looking at other cultures makes it apparent that the development of memory skills depends very much on cultural conditions—rudimentary skills as commonplace as rehearsal and hierarchical grouping may fail to develop under some conditions. People in different societies differ in their use of memory strategies, and these differences help determine how well they recall various types of information.

Some of the earliest work in this are was carried out by Michael Cole and his associates (Cole, Gay, Glick, & Sharp, 1971; Cole & Scribner, 1973). These researchers studied the development of memory among rural tribal people in Liberia. They found that memory skills, such as *categorical grouping* during free recall, occurred only when the Liberian children attended school. Those who did not attend school did not group items by categories during free recall. This occurred even though the schools attended in these studies were typically poorly equipped, one-room sheds, stressing methods based on rote memory.

Several studies by Wagner (1974, 1978) showed that schooling also tends to promote *rehearsal.* Wagner studied Mayans living in the Yucatán and later schooled and nonschooled children and adults living in Morocco. In the latter study, he found that children who attended school began to use rehearsal but noneducated Moroccians did not. Again, schooling appears to be the critical variable. The development of memory strategies such as categorical grouping and rehearsal seems to be a direct result of having to learn fragmented bits and pieces of information that initially lack any intrinsic and coherent structure.

Such skills are unnecessary in societies without formal education and are, therefore, unlikely to develop. Much of the education in these societies is informal and occurs during the course of day-to-day activities (Scribner & Cole, 1973). Such education tends to be highly particular, traditional, socially relevant, and fused with emotional content. It also tends to be based on observational learning and demonstration rather than description. Learning in school relies more on verbal description of unknown people, places, and events. Much of this material occurs out of context as isolated bits and pieces of information. Formal education and literacy encourage a reflective attitude toward language and memory that tends to promote the development of metacognition and memory skills. The end result is that literacy tends to greatly expand semantic memory at the expense of the episodic, and it also leads to the development of conscious strategies for enhancing the recall of a seemingly endless flow of information.

It would be wrong, however, to assume that people who have been formally educated have *better* memories. They perform better on standard memory tasks, such as list learning, paired associations, and free recall, typically used to study memory in experimental situations; but they perform significantly worse on other memory tasks more common in day-to-day life. Kearins (1981), for example, compared the spatial memory of Aboriginal and Anglo-Australian children by showing them a number of items arranged in matrices for 30 seconds and later asking them to replace them. She controlled for familiarity by using both natural and manmade objects. The Aboriginal children were consistently better at reproducing the displays, regardless of the number or items, the degree of similarity, or the type of material. Similar (although nonsignificant) differences were found by Rogoff and Waddell (1982) in a comparison of Mayan children and those living in the United States. Their data suggest that the Mayan children may have used a spatial strategy to remember objects, whereas American children sometimes used rehearsal. Rehearsal is not a very good strategy for visual displays because correct replacement of items depends on remembering the place and orientation of objects rather than the name.

Dubé (1982) has found that both literate and nonliterate Africans recall stories better than their American counterparts. When these three groups were separated on the basis of "intelligence" (using triable ratings and grade-point averages), the most intelligent nonliterates outperformed all other groups, and the most intelligent Americans performed about as well as the least intelligent Africans. These differences can be explained by the fact that African cultures encourage storytelling whereas American culture does not. Other examples of quite remarkable feats of memory in preliterate and nontechnological societies are described in Neisser (1982).

The question of who has the better memory may simply be the wrong question. People in all cultures develop memory strategies in an effort to cope with the pressures of their environment. The importance of these differences should not be underestimated, however. Success in a highly technological society often depends on the ability to learn numerous isolated items of information that initially lack any coherent organization. Courses in introductory psychology, for example, appear very different to teachers and students. For students, these courses seem to include a great deal of unrelated information that must be memorized and recognized on multiple-choice exams. Teachers see a much tighter organization. Nevertheless, almost all introductory psychology instructors were once students in similar courses, and these classes provided them with the background needed to select upper-level courses and go on and become psychologists. As Flavell (1978) has pointed out, effective performance in our culture often depends on the development of memory strategies and skills that allow people to remember uninteresting and initially meaningless material.

Other societies have developed different and sometimes quite elaborate *mnemonic strategies.* The Greeks and Romans, for example, developed extensive bureaucracies to maintain and administer their empires, but they used relatively little paper or written material. They were, therefore, more or less forced to develop techniques for remembering things without writing them down. One of their devices was the *method of loci.* Orators and those involved in the administration of public affairs would select a large building or structure and commit it to memory. They would then fill their imaginary space with mental images of topics to be remembered and recall the material by taking an imaginary stroll through the artificial space.

Another technique involves the use of *peg words.* A person begins by memorizing a poem, such as:

> One is a bun; two is a shoe;
> Three is a tree; four is a door;
> Five is a hive; six is sticks;
> Seven is heaven; eight is a gate;
> Nine is wine; ten is a hen.

A list of grocery items, such as eggs, milk, jam, and bacon, can then be remembered by forming mental images that link each word with the object in the poem—a bun with eggs, a shoe with milk, a tree with jam and a door with bacon.

A third technique uses *key words.* Some materials, such as foreign words, are difficult to learn because there is often no apparent link between the vocabularies of different languages. There does not, for example, seem to be any connection between the Spanish words, *clavo* and *pato,* and their English equivalents, *nail* and *duck.* In the key-word technique, a related word is found for each of the foreign words, such as *claw* and *pot* and these are then linked to the English word through a mental image. One could imagine a claw being driven into a board with a hammer, for example, or a duck with a pot on its head.

These and other artificial techniques are usually at the root of techniques designed to build a "super-power" memory. They also appear to have undergone periods of popularity and decline and have been reinvented from time to time by people with special needs. The use of mnemonic strategies became less necessary after the fall of Rome but they were revived in the Middle Ages. Access to cheap paper and the later development of the printing press reduced the need for artificial strategies. Somewhat later mnemonic strategies came under more direct attack by the English Puritans who objected to the techniques because they felt that mental images excited the imagination. They stressed rote memory instead (see Bolles, 1988). Mnemonic strategies based on visual imagery are not widely used in Western cultures, even when they are known. People needing items at a supermarket, for example, are far more likely to jot them down than take an imaginary stroll through their home. Other cultures have developed additional techniques to aid memory, and there appears to be a great deal of commonality among cultures widely separated in both space and time. Those with a strong oral tradition often convey fables and legends through poetry and song. The rhythm and lyrical quality of these media lend themselves to easy recall.

Many of the deliberate strategies people use to improve memory are acquired cognitive skills. They develop gradually over time and may fail to occur if they are not called on or used. If the demands on memory are great, this may mean the development of highly elaborate and sophisticated strategies. The high frequency of similar skills among adults in our culture has misled many people into assuming that they are an inherent part of information processing based on the architecture of the brain. This is true to a certain extent. We all have a tendency to organize and chunk information and make it personally relevant by relating it to what we already know. But the ability to use other more self-consciously directed skills may not occur if they are not needed. As Wagner (1978) has shown, even a technique as elementary as rehearsal may not be widely used when it is not required.

The changes that take place in memory occur as people adapt themselves to different environments. The demands placed on memory in nonliterate societies with a strong oral tradition are quite different from those in advanced industrial societies with ready access to paper and computers. Most people seem to have the ability to rise to the occasion and develop the skills necessary to get by on a day-to-day basis. Those who do not are severely handicapped because many of the higher order mental operations, such as problem solving and formal thought, depend on memory.

CHAPTER 10

Creative Problem Solving

The chapter begins by making a distinction between formal problems, which contain all the information needed to reach a solution, and informal problems, which require creative solutions. Gestalt and contemporary approaches to problem solving are discussed and used to support a model for insight and problem solving that combines features of each. The concept and childhood antecedents of creativity are described, and the final section deals with the more general concept of "practical intelligence." Throughout the chapter, it is argued that creativity and practical intelligence require unique skills that are unrelated to academic ability and, therefore, are not accurately assessed by traditional measures of I.Q.

American pragmatists, such as Charles S. Peirce (1878/1955), William James (1890), John Dewey (1911/1933), and George Herbert Mead (1934/1962), saw problem solving as the source of most conscious behavior (see Collier, Minton & Reynolds, 1991). Consciousness occurs during the course of concrete activity when there is a decision to be made or a problem to be solved. Once a person has acquired new behaviors, they gradually become more habitual and recede from conscious awareness. Habits simplify behavior, increase accuracy, and diminish fatigue. Previous goals become means for new goals, and each person builds up a repertoire of habitual responses that function smoothly and effortlessly in most situations.

Purposeful activity occurs when these habitual responses are no longer appropriate. People begin to selectively attend to new features of the environment and imagine alternative ways of reaching the same goal. A complete cycle of activity ends in behavior. Deliberate behavior need not be completely thought out, however. All that is needed is a general notion of a particular goal and several ways of reaching it. The pragmatists were concerned primarily with concrete thinking and problem solving and regarded speculative thinking, daydreaming, and other forms of "private" thought as a derivative. Concrete thinking does not just occur "within" the head of an individual—it draws from concrete objects and situations in the outside world.

Problem solving may involve *formal* or *informal* reasoning (see Galotti, 1989). Formal problems such as logical problems, mathematics, most analogies, and series completion contain all the information needed to reach a solution. Formal problems are frequently used in laboratory studies of problem solving because they have one and only one correct answer and are, therefore, easy to designate as right or wrong. They are also included on various intelligence tests because the solutions do not depend on previous experience or memory. All the premises and information necessary to reach a solution are explicitly stated so that people do not have to rely on memory or consult other sources to obtain information.

Informal thinking occurs when problems are not well defined. Most of the problems in everyday life are ill defined and require people to seek out new information and consider alternatives to reach a solution. Others require a great deal of research or "soul searching" before they can be resolved. Informal problem solving also differs from formal reasoning in that there is no one correct solution. Answers to formal problems (such as $2 + 2 = 4$) are either right or wrong, whereas answers to informal problems may be better or worse. Solutions to everyday problems may have both good and bad features, depending on a person's perspective, and it is always possible that a better or more creative solution could be found if more information was available.

Various people have attempted to describe the relationship between formal and informal problem solving and Galotti (1989) has summarized three different positions that have been adopted. Some, such as Halpern (1984) and Wason and Johnson-Laird (1972), view formal reasoning as a part of everyday reasoning. Everyday reasoning requires an additional step because people must generate their own premises but, once the premises have been generated, the reasoning process is more or less the same. Implicit in this position is the belief that everyday reasoning is more difficult because it involves additional steps and often various forms of emotional commitments and preconceptions which must be overcome.

A second view, which is in many ways to the opposite of the first, is that both formal and informal problem solving use similar processes but formal reasoning is more difficult because people must selectively focus on the premises themselves and ignore related information based on personal associations and memory. Some problems in logic, for example, are extremely difficult because they fly in the face of common sense. Many people have difficulty with syllogisms, such as,

All birds are turkeys.
Penguins are birds.
Therefore, penguins are turkeys.

Yet, the logical structure is precisely the same as the well-known syllogism, "All men are mortal. Socrates is a man. Therefore, Socrates is mortal." The inability to strip away nonessential features makes formal logic difficult, and overcoming this handicap often requires considerable exposure to logical problems.

A third position is that formal and informal problem solving are simply different (e.g., Perkins, 1986). Formal reasoning has a "long chain" structure with many different steps each leading to the next one in the sequence. Informal problem solving has a "forked" structure that requires decision making from time to time, and there are frequently many different ways to reach the same conclusion. Formal arguments are either true or false, whereas solutions to informal problems typically include both good and bad features so that some decision about the relative merit of each must be made. Formal problem solving occurs in a closed world that encompasses all the information in advance, whereas informal problem solving draws from other sources, such as previous experience, outside sources, and creative imagination.

Because formal and informal problem solving require different cognitive skills, it makes little sense to argue that one is inherently more difficult than the other. Some formal problems in logic and mathematics are so difficult that they elude all but the most gifted thinkers. Other informal problems in science are so inherently complex that they seem to defy any solution given our current state of knowledge. The concept of creative problem solving will be the focus of the present chapter, whereas the concept of formal thinking (broadly defined) will be deferred until Chapter 11.

PROBLEM-SOLVING APPROACHES

Gestalt psychologists were among the first to attempt to explore processes involved in problem solving. Although they are best known for their work in perception, they also carried out extensive research on learning and problem solving and were vocal critics of both introspection and American behaviorism. For Gestalt psychologists, problem solving involves a recognition of the structural relationships between various elements of a problem and a restructuring of the psychological field to reach a solution.

One of the central concepts of Gestalt psychology is the notion of *insight,* which involves a sudden grasp of how individual elements can be reorganized to solve a problem. The best known illustrations of insight learning are Wolfgang Köhler's studies with chimpanzees on the Isle of Teneriffe in the Canary Islands. Köhler, a German psychologist, went to a research station on Teneriffe in 1913 and was trapped there during World War I. Altogether, he spent 7 years on the island carrying out various studies of insight learning. In one study, which was previously mentioned, a chimp discovered he could stack several boxes to reach a banana that had been suspended from the roof of his cage. Using this procedure, one chimp could build rickety structures up to four boxes high. In a second study, a banana was placed outside a cage with a shrub inside. Sultan, whom Köhler described as his "genius" chimpanzee, solved the problem by breaking off a branch, tearing off its leaves, and using it as a stick to rake in the banana.

In a third study, a banana was placed outside the cage and two sticks were placed inside. Each stick was too short to reach the banana but they could be fitted together to make one stick of sufficient length. While playing with the sticks, Sultan noticed that the smaller one could be inserted into the larger one. The chimp then put them together and immediately used them to draw in the banana. Köhler's findings were published in German in 1917 and were later translated into English and published under the title, *The Mentality of Apes* (Köhler, 1925).

As mentioned in Chapter 7, the term "insight" was deliberately chosen to stress the visual nature of the solution. Köhler's chimps could solve a variety of problems if all the necessary elements were present in the visual field and sometimes when they were not. They could, for example, turn, go out a door, run down a hallway and go outside to get a banana they had previously seen through a window, and they would dig up fruit that they had seen buried up to four days before. But most of their solutions were "visual" solutions and rearrangements of objects which were visually present. Behaviorist critics argued that the solutions involved a great deal of trial and error, but the quickness of the response after the relationships were noticed suggests that the solutions occurred suddenly as a single burst of insight. Chimps, who are incredibly expressive, showed their insight by hesitating, taking a survey of the situation, and then responding quickly after the solution was found. Insight learning is often described as a *Eureka!* experience in which the solution to a problem is suddenly grasped as a meaningful whole in a single instance.

In humans, insight is not limited to visual problems but the phenomenon shares many of the same features. Insights can occur in any area, including abstract reasoning and formal thought. One of the best examples of insight is the recognition of an embedded figure in a complex design. When someone looks at the drawing initially, there is no recognition, but eventually the embedded figure begins to stand out. The moment of insight is like a burst of illumination in which a previously hidden figure suddenly becomes visible.

Insights are often hampered, however, by *problem-solving sets,* which are habitual ways of solving problems based on familiar or routine solutions. Once a solution has been found for a problem, it is often used to deal with similar problems and becomes more and more automatic. Such habitual use of routine solutions makes day-to-day activity easier but it often prevents better and more creative solutions. People are unable to break away from traditional solutions and see similar problems from a fresh perspective.

Somewhat related is the concept of *functional fixedness,* which is a tendency to see objects in terms of their conventional function. This concept can be illustrated by an experiment carried out by Karl Duncker (1945), which is known as the "candle" problem. Duncker gave subjects a candle, a book of matches, and a box of thumbtacks and asked them to mount the candle to the wall so that no wax would drip on the floor. The solution involves using the box as a holder for the candle and mounting it to the wall with the thumbtacks. When the box was presented with the thumbtacks inside, 80% to 90% of the subjects failed to solve the problem because they could not disassociate the box from its function as a container. When an empty box was presented, however, 80% of the subjects found the solution.

Once people break free of functional fixedness and solve a problem, they are often astonished at how easy the solution is and how much time it took them to discover it. This explains why many groundbreaking scientific discoveries seem surprisingly simple in retrospect and why they are often made by people coming from slightly different areas of concentration. A strong indoctrination within a particular discipline often leads to functional fixedness, and those with less indoctrination or other outside interest often see that discipline from a different perspective (Kuhn, 1970).

One of the problems with the Gestalt approach is that, although it identified a phenomenon that occurs frequently, it did not provide much insight into the nature of the process. Insight probably occurs because a number of processes occur more or less automatically without awareness. What the person experiences is a quick transition whereby a problem that previously had no solution suddenly has one. Problem-solving sets or functional fixedness can hamper the process, but there is little indication why insights occur or fail to occur or why some people are more prone to them than others.

Both Walps (1926) and Hadamard (1949) have provided a more comprehensive description of insight by placing problem solving within the context of four stages—preparation, incubation, illumination, and verification—in which the third stage is the moment of insight. *Preparation* involves an intense effort to find a solution to a particular problem, which may or may not lead to a suitable response. If the problem is particularly difficult, people eventually give up and enter the stage of *incubation,* during which their conscious attention is directed toward other problems. Incubation may take anywhere from a few seconds to a few years, depending on the problem, but eventually the solution "processes itself" and the person experiences a sudden bolt of *illumination,* which characterizes the *Aha!* or *Eureka!* experience. After this, the person must check the validity of the insight by examining the details during the process of *verification.*

These theories involve a blend of conscious and unconscious processes, plus what William James (1890) labeled the "fringe consciousness" and some psychoanalytic writers (e.g., Kubie, 1958) call the "preconscious." Preparation and verification rely on conscious processes that are deliberately used to solve a particular problem. The quality and nature of the solution depend on how much time and effort a person is willing to expend during preparation and verification. Unconscious processes not available to introspection occur during the stage of incubation when the mind is occupied with other things. It is these processes that give insight its "mystical" quality and dissuade some from doing research on creativity. Fringe consciousness, which occurs during illumination, is the grey area between consciousness and unconsciousness. Insight occurs as ideas move from the unconscious to the fringe consciousness and then are brought into full awareness.

This moment of insight can be better understood by looking at each stage in terms of more contemporary models of information processing. When someone begins to study an area, such as the various approaches to problem solving, he or she is initially struck by a bewildering number of competing descriptions, each describing a similar process in slightly different terms (see Galotti, 1989). The more the person learns, the more the similarities begin to stand out and the differences begin to fade. The learner begins to abstract an underlying pattern common to many different theories. This pattern resembles a "composite picture" made up of numerous superimposed images of a similar class of objects.

During this process, perception and memory are being modified in ways described previously. The individual becomes progressively better at selectively attending to important details and ignoring the less relevant. Information processing occurs more quickly and is organized into increasingly larger chunks of information. These chunks themselves become associated with other chunks in a hierarchical organization. If information processing proceeds smoothly, then a *pattern* eventually emerges or "falls out," which is similar to recognizing an embedded figure in a complex design. It should be noted that a great deal of what is eventually recognized as creative problem solving occurs in this rather straightforward way without hesitation or delays. Ideas flow, patterns emerge, and understanding occurs more quickly, more easily, and in a more organized form.

If a problem is particularly difficult, on the other hand, then a pattern may not "fall out," and a solution may not be immediately obvious. More work simply delays the process, but the person eventually tires, gives up, and shifts attention to other matters. During this time, which may last a few seconds or a few years, the brain is busy unconsciously (and sometimes consciously) consolidating material. Details begin to fade and overlearned essential features begin to take the foreground. Eventually, a pattern begins to stand out, leading to a delayed recognition or solution to a problem.

This moment of insight shows that it is possible to know too much—to be unable to recognize the forest for the trees because of being overwhelmed with minute details. This occurs frequently in school, for example, where the details must be memorized and students are tested later. Forcing psychology students to memorize all the various aspects of Sternberg's (1988) triarchic theory of problem solving; Newell's (1980) problem-solving space; or the various componential, rule/heuristics, and mental models/search approach models described by Galotti (1989) provides them with the necessary background but also leaves them with a bewildering array of disconnected bits and pieces that contain all the information in the embedded pattern plus much, much more.

In other cases, the pattern does not fall out by itself. Instead, we are suddenly exposed to an analogy or metaphor that resembles the pattern in some way and the resemblance triggers the insight. Examples in the history of science are common. A famous example is Friedrich August von Kekulé's report of his discovery of the ring structure of benzene. According to Kekulé, he had been working on the problem for some time, and one night when he was dozing in front of a fire, he saw an image of a snake twisting around and seizing its own tail. In a sudden flash, he realized that the molecule must be structured as a ring. Such analogies can be based on physical events or on images occurring in dreams and daydreams. Those who treat dreams as random events would argue that, during the dream, the person is exposed to a variety of images and eventually finds a match. Kekulé, for example, (if he was dreaming) might have seen a house, a garden, a tree, and then a snake, but only the snake triggered the solution to his problem. Others, such as Freud (1900/1953), view dreams as the "royal road to the unconscious" and dream symbols as a means for making unconscious ideas known. In either case, the underlying similarity occurs as a sudden jolt of insight or recognition.

As Hadamard (1949) pointed out, the moment of insight is almost invariably followed by a period of *verification* because insights almost never occur in a form sufficiently clear to be expressed without elaboration. There are also many "false insights" which must be checked and eliminated, or partial insights which must be linked with other material and expanded enormously. The verification process may take moments or years.

During this process, metaphors take on a second role—that of *expression.* William James (1890), for example, used a variety of metaphors to describe the "stream of consciousness," including a train (which has its own power and can change speeds), a path, and hydraulic and electrical currents. Osowski (cited in Gruber & Davis, 1988) argues that James was unable to find a single unified image that captured both the phenomenological and neural quality of the stream of consciousness. Other examples include Darwin's (1859) use of a "branching tree" diagram to depict evolution (which is the only illustration in *The Origin of Species*) and Freud's (1915/1957) hydraulic model of motivation. It would be wrong to confuse these metaphors with the insights that gave rise to them. Because they are based on a deliberate search for the most appropriate means of expression, they both capture and gloss over aspects of the original understanding.

The preceding description attempts to combine the insights of the Gestalt approach with our more recent knowledge of how information processing takes place and "demystify" the creative process. Creative problem solving involves a combination of conscious knowledge and skills, plus a number of automatic processes, such as chunking, consolidation, and selective forgetting, which operate below the level of awareness. It sometimes involves an extensive period of preparation, which may include years of formal education. This provides the necessary background knowledge to recognize insights when they occur. Many people dream of snakes but only Kekulé saw the snake as a symbol for the structure of benzene. Creative problem solving also typically involves a period of verification, in which insights are tested, elaborated, and transformed into a form that can be communicated to others.

If this account is correct, then it helps to explain some of the controversies that have plagued the areas of problem solving and creativity. First, as mentioned previously, insight is *both* a conscious and an unconscious process. But the unconscious aspects of insight are no longer a mystery. They obey what are now well-accepted laws of learning and forgetting. Patterns emerge after long delays because overlearned, organized material is retained, whereas poorly learned, disorganized material is soon forgotten. Second, insight is *not* a necessary part of the creative process. It occurs only when there is a period of delayed recognition. Problem solving, when it proceeds smoothly, may involve a prolonged period of creativity without a moments hesitation. An author writing a novel, a painter choosing the next brush stroke or a scientist coming to grips with a complex conceptual problem may experience a continuous flow of ideas. Insights occur only when this flow is interrupted and people are unable to come to an immediate solution. Third, although metaphors are often cited as a source of inspiration, they too are *not* a necessary part of creative problem solving. Normal changes taking place during information processing, such as chunking, consolidation, and selective forgetting, allow insights to occur with and without metaphors. Metaphors can serve as a catalyst for delayed recognition or a means for expressing insights that have already occurred, but they are not absolutely necessary.

Finally, it is now possible to see that creative problem solving does *not* take place "in the head" of an individual. Insights are impossible without someone who can recognize them, but the pattern a scholar sees after reading and rereading countless articles is contained in the original descriptions. This is why they can be communicated and recognized by others with a similar background and training. Some of the most groundbreaking scientific discoveries seem surprisingly simple in retrospect. When T. H. Huxley first heard of Darwin's theory of evolution in 1858, his response was "How stupid not to have thought of that!" (Huxley, 1935). The simplicity of these ideas occurs only because they have been recognized and expressed, and they are, of course, not simple to everyone. Darwin's theory was not universally accepted even by the scientists of his day. Older scientists, in particular, had difficulty coming to grips with it and its implications. Laypeople were often hostile because they were misled by a religious establishment that saw the conflict between evolutionary theory and a literal interpretation of the Bible. Although it would be misleading to overestimate the consensus among informed observers, all great theories with personal or social consequences—Darwinism, Marxism, Freudian psychology, and so on—have been distorted in a way that makes them difficult to grasp and easily rejected by those with little knowledge.

What is true of scientists and scholars is true of artists and writers as well. No one starts in an intellectual or cultural vacuum. The artistic process often flows smoothly without hesitation, but when problems occur, they are often followed by moments of insight and periods of verification. Artists try out new modes of expression, reject them, and try again. Picasso, for example, worked many months on his gigantic mural *Guernica,* making sketches and revisions as he went along, and drew from both his own earlier work and that of other artists (Arnheim, 1962). Brahms took 20 years to write his first symphony because he was living in the shadow of Beethoven. Artists and writers are not always aware of the influence that others have on them. What seems to be an original idea may be a composite derived from many other people.

An adequate description of problem solving must take into account not merely the mental processes or strategies that people use in dealing with new situations but attitudes and predispositions that allow people to overcome problem-solving sets or to seek out information challenging their previous preconceptions. It must be able to explain why some people tolerate or even enjoy ambiguity, whereas others strive for solutions as quickly as possible. Or why some people are willing to think "long and hard" about a particular problem, returning to it again and again, over a period of weeks, months or even years. As Galotti (1989) points out, some of what predicts good problem solving is based on the depth and breadth of the knowledge base and the amount of previous experience, but good everyday problem solving also depends on a more thorough and less biased exploration of counterarguments and evidence and a more sophisticated use of existing information.

Because the focus of the present chapter is on creative problem solving, the concept of creativity will be discussed briefly in the next section. The research on problem solving and on creativity have been conducted, until recently, in virtual isolation. Research on problem solving tends to focus on the processes involved in solving problems, whereas that on creativity focuses on traits characteristic of creative and noncreative people. The section on creativity will be followed first by a review of the childhood antecedents associated with creativity and then by a more general discussion of practical intelligence.

CREATIVITY

Most everyday problems are ill defined and therefore require creative solutions. Almost by definition, people lack the resources to deal with the problem immediately and must generate possible solutions by searching their memory or seeking additional information. It is useful to break down this process into two sets of skills. First, people must generate a number of possible solutions; second, these must be evaluated and compared so that the best one can be selected. The need to generate possible solutions is what distinguishes well-defined and ill-defined problems and the ability to do this varies from person to person.

These two aspects of creative problem-solving correspond, more or less, to Guilford's (1967) distinction between divergent and convergent thinking. Convergent thinking occurs when there is one correct solution to a problem. The thought processes *converge* on the right answer. These are the kinds of problems that typically occur on intelligence tests. Divergent thinking is based on the ability to produce many different alternatives for problems that have more than one solution. These differences can be illustrated, using the example of the logical syllogism described previously (for convergent thinking) and "How many uses can you think of for a brick?" (for divergent thinking),

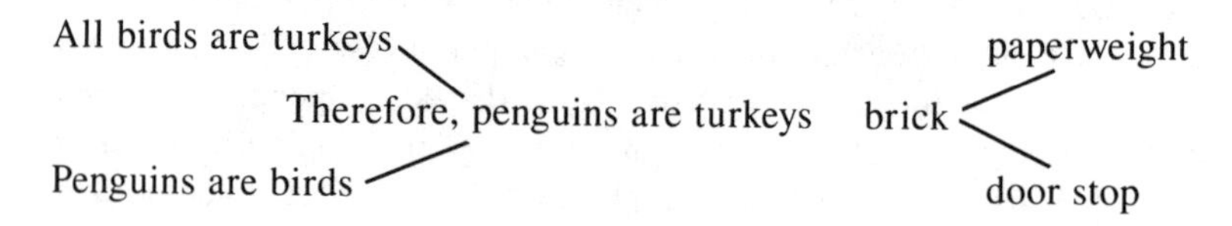

In the case of the logical syllogism, there is one and only one correct conclusion, whereas there are almost an endless number of things you can do with a brick.

Guilford (1967) argues that divergent thinking is based on four skills—fluency, flexibility, originality, and elaboration. Fluency is the ability to generate a number of responses, whereas flexibility is the ability to change the type of response. If someone says that a brick can be used to build a house, a barn, a school, a store, and so on they would score relatively high on fluency but low on flexibility. Another who says that a brick can be used to build a house, drown a cat, make bookends, or mark the grave of a small bird would score high on both. Originality has to do with the uniqueness or quality of the responses, whereas elaboration is the ability to expand ideas or fill them out in detail.

Although some claim that there is a strong link between convergent and divergent thinking, most researchers have found little or no relationship. Guilford (1967) argues that there is a curvilinear relationship between convergent and divergent thinking. Those with low IQs seldom score high on measures of divergent thinking, whereas those with high IQs may be high or low. Simonton (1984) suggests that there is a positive relationship between creativity and IQ but that this relationship seems to vanish somewhere above 120. Beyond this point further increases in IQ do not increase the likelihood that someone will be more creative. An IQ of 120 is not very high. It is the average IQ of college graduates and about 10% of the general population have IQs above 120. Tests comparing highly creative people with less creative people in the same field have found little relationship between convergent and divergent thinking. Exceptionally creative architects, mathematicians, and scientists, for example, usually score no better on intelligence tests than their less creative peers (MacKinnon, 1978).

The lack of correspondence between traditional measures of intelligence and creativity can also be seen in the recent research on *birth order.* It has long been known that firstborn siblings tend to score higher on standard intelligence tests than only children or younger siblings (e.g., Zajonc & Markus, 1975; Zajonc, Markus, & Markus, 1979). They also make up a disproportionate number of National Merit Scholarship winners. Because these results cannot be due to genetics, they are often explained in terms of interactions and relationships within the home. Firstborn children receive more attention from the parents, whereas later-born children must share the parents' attention. Firstborn children also instruct their younger siblings and, thus, develop the reflective skills that occur commonly on IQ tests.

Research has also shown, however, that political revolutionaries are much less likely to be firstborn (e.g., Stewart, 1977; Walberg, Rasher, & Parkerson, 1980), and Sulloway (cited in Woodward & Denworth, 1990) has recently argued that really important "scientific revolutions" are typically the product of younger siblings. He suggests that firstborn children are more likely to identify with their parents and adopt the norms and values of society, whereas later-born children are more likely to question these values and strike out on their own. Although the effects of birth order on both IQ and creativity are extremely weak, they suggest that the two should be kept separate.

Tests of divergent thinking, on the other hand, have been linked to creativity. Wallbrown and Heulsman (1975) gave tests of divergent thinking to children in the third and fourth grade and found a significant relationship between these scores and the ability to create original clay sculptures. Other research has found a small but significant correlation between divergent thinking scores and such activities as creative writing, scientific experiments, and drama (Rotter, Langland, & Berger, 1971). Research by Torrance (1972, 1975) suggests that performance on creativity tests given in high school can predict the level of creativity up to 12 years later.

Creativity has been linked to a number of other traits as well. These include a tolerance for ambiguity, a lack of functional fixation, stimulus freedom, and risk taking. *Tolerance for ambiguity* is often seen as one of the hallmarks of creativity (Barron, 1968; Torrance, 1979). The Barron-Welsh Figure Preference Test, for example, was developed to measure people's preference for simple and symmetrical figures or complex and asymmetrical ones (Barron & Welsh, 1952). Barron (1968), who has carried out a series of studies of people in various professions, including artists, architects, and scientists, has found that those who were selected as creative by their peers tend to prefer more complex patterns. He describes this as a *preference for disorder.* Such pictures are more interesting not because they are disorganized but because they are more challenging. Creative people are capable of tolerating and even enjoying a certain amount of disorder because it allows them to organize the material in their own way—bringing order out of chaos.

Creative people are also more *open-minded* and are less likely to experience functional fixation. They respond to new information without prejudice and are able to imagine alternative uses for familiar objects. Rokeach (1960) has reviewed the literature on open-mindedness and has concluded that open-minded people are less rigid, less neurotic, less anxious, and less authoritarian. The authoritarian connection is important because it links creativity or, more accurately, noncreativity to a particular child-rearing pattern and helps to explain why lack of creativity is so common. MacKinnon (1962) has found that creative architects were more flexible and open-minded and had a wider range of interests than noncreative architects (see Barron & Harrington, 1981).

Somewhat related to open-mindedness is the concept of *stimulus freedom.* Stimulus freedom has two aspects. First, creative people are more likely to bend the rules when they interfere with the creative process. They are less constrained by conventional standards and norms and more likely to break rules if it seems necessary. Stimulus-bound individuals follow rules to the letter and use them to guide behavior. Second, creative people are less likely to assume rules exist when they are not explicitly stated (Getzels, 1975; Taylor, 1975; Torrance, 1979). The nine-dot problem, for example, consists of three rows with three dots in each one which are to be connected by four straight lines (Figure 10.1). The solution (Figures 10.2) requires that people leave the boundaries of the figure, but many people have difficulty doing this because they implicitly assume that the solution must be contained within the figure itself.

A final factor associated with creativity appears to be *risk taking.* Creative people are more likely to take risks and can go for longer periods without receiving rewards. Vincent Van Gogh, for example, sold only one painting in his own lifetime (and that to his own brother), yet he knew his work was good and approached it with a burning obsession. Creative people are less restrained by a fear of the unknown or a fear of failure. They generally select tasks that are moderately difficult, rather than extremely difficult or extremely easy. They are also less concerned about what others think and are more likely to judge their work by their own standards. The concept of fear of failure and its impact on performance has been discussed previously. What is important here is to note that this traitlike characteristic can interfere with the creative process, and like the other traits mentioned previously, it has its roots in childhood and culture.

Figure 10.1. The nine-dot problem. Connect all the dots with four straight lines. (The solution is given on page 196.)

CHILDHOOD ANTECEDENTS

The ability to approach problems in a novel and creative way, like most skills discussed in this book, seems to have its roots in early childhood. Each of the various attempts to explore these childhood antecedents confronts a particular set of problems. If the researcher begins with children, for example, measures of creativity are not necessarily good predictors of actual performance. Significant correlations have been found but these cannot be used to determine whether one child or another will go on to be truly creative later in life. A researcher who starts at the other end and examines the lives of people who have achieved eminence (e.g., Goertzel, Goertzel, & Goertzel, 1978; Simonton, 1984) or who are currently rated highly by others in the same profession (e.g., MacKinnon, 1978) is limited to biographical material or retrospective reports of early childhood, which are equally dubious and subject to distortions. A middle position looks at creative adolescents—that is, high school and university students who display high levels of creativity (e.g., Dacey, 1989), and this approach combines both the strengths and weakness of the previous two. Despite the lack of an ideal method, certain patterns have emerged.

One common pattern found with a variety of different procedures is that creative people come from homes in which they were given or allowed a great deal of *freedom and autonomy,* either because they were neglected or permitted to decide things for themselves. Parents with creative teenagers tend to be less authoritarian and impose fewer explicit rules, such as when to study or go to bed or how late to stay out at night (Dacey, 1989). Instead, such parents set general guidelines within which children decide things for themselves. These flexible guidelines and lack of strict

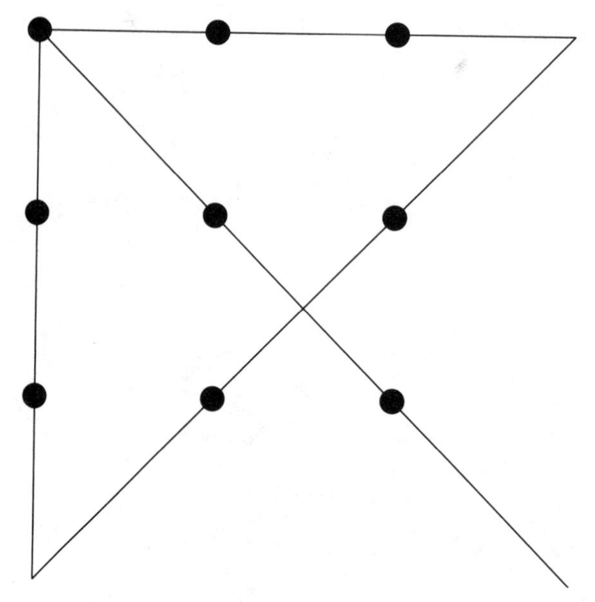

Figure 10.2. Solution to the nine-dot problem.

discipline seem to promote a more flexible and open-minded attitude that encourages creativity.

Observation of parents and children in the home suggests that a second factor associated with creativity is *humor* (Dacey, 1989; Torrance, 1979). Parents with creative children tend to engage in a great deal of joking and fooling around. These studies suggest that a playful attitude is an important part of the creative process. Other studies have shown that preschool children who engage in a great deal of fantasy score higher on measures of divergent thinking (Dansky, 1980; Johnson, 1976). Playfulness and fantasy seem to work against functional fixity and encourage a wider range of creative options.

A third positive source of inspiration is an *early love of learning*. Goertzel, Goertzel, and Goertzel (1978) examined the lives of 317 "eminent" people—that is, writers, artists, and other famous people living in this century and born in the United States who have been the subject of at least two biographies and those born outside the United States who have had at least one biography. They found that almost all the people in their study (90%) came from homes in which there was a love of learning by one or both parents, and this quality was often shared by grandparents and siblings. Half of those studied were early and eager readers and they continued to read voraciously throughout childhood and later life. An early love of learning seems to be a necessary part of creative people's development because it provides them with the necessary background and special skills needed to excel later.

Despite these auspicious beginnings, many eminent people, particularly, writers and artists, came from homes in which they were *desperately unhappy* (Goertzel, Goertzel, & Goertzel, 1978). Only 28% of the people in the Goertzels' study described their homes as "usually quite happy." Some of the problems, such as wars, discrimination, and political changes, occurred because of events outside home. Others, such as physical illnesses, trouble in school, quarreling parents, deaths, desertions, and financial ups and downs, occurred within the family. One in four experienced serious or chronic physical illness during childhood, and the authors speculate that this early illness took them away from other activities, forced them to turn inward, and thus helped promote the introspective skills common among creative writers and artists. In 40% of the homes, the father was a poor or erratic provider, and these families often experienced vacillating periods of wealth and poverty. Very few, however (only 6%), came from homes in which poverty was the norm.

MacKinnon (1978) and his associates have studied creativity more directly by comparing creative writers, architects, mathematicians, scientists, and engineers rated highly creative by those in their profession with less creative people in the same field. They have found a similar pattern, in which creative people were more likely to describe their childhoods as unhappy. Although their parents granted them a great deal of freedom, there was typically a lack of closeness on the part of one or both parents. Many of the creative adolescents in Dacey's (1989) study also suffered a large number of *traumatic experiences,* that caused grief or seriously disrupted their life.

One particularly common hardship among creative people is *the loss of a parent.* The most systematic study of parental loss was conducted by Eisenstadt (1978) who examined the lives of 699 eminent people from various walks of life and found that almost half had lost one or both parents before the age of 21. The most common loss was the death of a father by a creative male. This seems to encourage the development of an androgynous or somewhat feminine personality. MacKinnon's (1978) administered several personality inventories to contemporary people who were rated as highly creative in their field and found that one of the most striking features was that they scored extremely high on measures of "femininity" and possessed such traits as self-awareness and openness to their feelings, and displayed a wide range of interests that are typically considered to be feminine in our culture. In other cases, the absence of a strong male role model seems to have produced an exaggerated overcompensation, leading to the development of "macho" features that are commonly found among dictators, military leaders, and some artists and writers (Simonton, 1984).

The overall picture that emerges is quite different from the usual finding that positive traits, such as high levels of achievement motivation, an internal locus of control, high levels of aspiration, and self-esteem, are generally promoted by warmth and affection within the home. Many of the creative or eminent people in these studies did receive warmth from at least one parent, but their childhoods were far from ideal. In many ways, they resemble the resilient children mentioned briefly in Chapter 4, who overcome hardships (with the help of others) and gain the independence and insight necessary to be truly creative. Psychologists are a long way from fully understanding why negative experiences and childhood traumas lead to a loss of control, feelings of helplessness, depression, and suicide, in some cases, and a sense of mission, increased productivity, and creativity in others. Writers, in particular, are often inspired by painful childhood events and use fiction as a way of recapturing and coming to grips with them.

A final feature characteristic of many creative people is a *negative attitude toward school.* Goertzel et al. (1978), for example, found that 60% of the eminent people they studied disliked school and had difficulty adjusting to teachers and peers. The effects of school on intellectual development have been described in more detail by Scribner and Cole (1973). They point out that, in many cultures, children are initiated into the adult world without formal instruction. They participate initially as spectators, with adults serving as models. As they become more and more experienced, they receive more and more responsibility. Adults and children soon begin to share the same work, with children gradually taking the initiative and adults correcting and guiding their behavior. Eventually, children are given most of the responsibility and adults become inactive but supportive spectators.

Under these conditions, learning occurs smoothly and effortlessly at a pace set by the children. Adults provide a comfortable degree of challenge that encourages growth and development. The group nature of most activities means that it is often difficult to separate individual contributions because everyone is participating at the same time. Children perform within their limited zone of competence, and they are rarely asked to perform beyond their level of ability. The group accepts the growing child's limited ability and does not openly expose their shortcomings and areas of ignorance.

Formal schooling, in contrast, places a heavy emphasis on individual performance and competition. Teachers provide information and directions within a structured situation, and children are expected to learn the material and repeat it back. In a typical classroom, teachers ask questions, children respond, and teachers evaluate these responses as right or wrong. Some children perform well, but a sizable minority react by becoming passive learners. Those who perform poorly develop a sense of personal incompetence and a devastating loss of self-worth and self-esteem. They may describe themselves as "not good in school" or "too stupid to learn" and develop the pattern associated with learned academic helplessness, described previously. They often come to devaluate academic tasks altogether and develop compensatory strategies designed to preserve their sense of self-worth and self-esteem.

Although school facilitates the development of complex skills, such as abstract thinking, hypothesis testing, and the deliberate use of conscious strategies, it appears to inhibit more spontaneous forms of search and discovery and more creative forms of problem solving. Children are increasingly exposed to problems with one and only one right answer. The authoritarian nature of the classroom tends to promote a passive attitude in which information is learned for no apparent purpose. For those who do poorly, academic success may lose any significance and outside activities come to be the central focus.

The one group which seems to be a general exception to the above description is *scientists* (Goertzel, Goertzel, & Goertzel, 1978). Scientists generally come from happy homes, do well in school, and later became loving and attentive parents, who occasionally neglect their families when overly involved in their own research. A career in science also involves a prolonged period of education and, typically, an apprenticeship under those already working in the field. This apprenticeship and the close one-on-one relationship that it encourages often determine the amount of success later in life. Zuckerman (1972), for example, found that over half of all Nobel laureates in science were apprentices to previous Nobel laureates.

Although childhood experiences seem to play a major role in the development of creativity, creativity varies in rather complex ways throughout life. Children who show few signs of creativity may later become creative if they find a nurturing and supportive mentor. Many creative people are very precocious, whereas others seem to get a late start in life. Several studies have examined the relationship between creativity and age. An early study by Lehman (1953) examined the biographies of several thousand people and compared the ages at which they made major contributions. He found that creativity increased throughout the 20s, peaked during the 30s, and then declined.

Dennis (1966) later pointed out a critical flaw in Lehman's research. Lehman's sample included a large number of people who died young and could therefore not make contributions later in life. Dennis studied 738 creative people who lived to be 79 or older and found little evidence for a decline in creativity as people grew older. Artists tended to increase their output throughout their 20s and 30s and peak during their 40s. Scholars and scientists, on the other hand, produced very little in their 20s and peaked between 40 and 60 years of age. Most produced as much during their 70s as they did early in life. Scholars and scientists require extensive training and a large knowledge base that makes later contributions easier. These appear to offset declines in physical energy, perceptual ability, and so on.

More recent research by Simonton (1984) suggests that quantity of creative output may decline but quality does not. He broke the creative process into two stages—ideation and elaboration. Ideation is the development of the initial idea, whereas elaboration is the working through of this idea until it becomes a finished product. In some areas, such as poetry, art, and certain forms of music, the length of time occurring between the initial insight and the finished product is relatively brief, and people in these fields tend to peak early in life. In other areas, such as history, philosophy, and, to a lesser extent, science, there is often a lengthy period of development and scholars in these areas tend to be quite productive well into old age.

One of the things that distinguishes creative and noncreative people is their *level of productivity.* Those rated as highly creative tend to possess high levels of energy and seem to have a sense of personal destiny that "drives" them to work hard and create (MacKinnon, 1978). They also find a great deal of intrinsic enjoyment in their work, so that working hard becomes a source of self-fulfillment and discovery. Simonton (1984) provides some examples of this relentless drive. Thomas Edison held 1,093 patents. Albert Einstein had 248 publications; Sigmund Freud had 330. Mozart composed well over 600 works before he died at the age of 35, and Schubert composed more than 500 works before dying of typhus at the age of 31. Bach averaged 20 pages of finished music per day. Rembrandt produced around 650 paintings, 200 etchings, and 2,000 drawings; and Picasso created more than 20,000 works of art. Although there are exceptions, there is a close connection between the quantity and quality of creative work. Those who achieve eminence typically start early, produce at a fabulous rate, and remain productive well into their old age. This high rate of productivity shows how important motivation is in the creative process and supports the famous quip by Thomas Edison, "Genius is one percent inspiration and ninety-nine percent perspiration."

In fact, this intrinsic interest and high level of productivity may be the fuel that feeds the creative process and, ultimately, may be more important than any amount of "inherent potential." By working hard, learning more, and developing special skills, people develop the self-confidence and ability to be more creative later. Creativity is a *circular reaction* that feeds on itself. Each creative act changes the person who initiates it—mildly for the most part but more radically in other cases.

Examination of the social origins of creativity raises an interesting paradox. Creativity, which seems to be the ultimate form of personal expression and free choice, is itself highly determined by social conditions over which children have little or no control. Early childhood experiences can instill a burning desire to excel or create a sense of helplessness that makes creativity virtually impossible. The ability to keep an open mind and pursue various alternatives is based, at least in part, on patterns of parental discipline. In later life, people are shaped by the broader social context and by the people with whom they happen to have firsthand experience. Both artistic and scientific discoveries are anchored in a reality that is not always chosen. Insights are often based on chance or fortuitous circumstances. Creativity lies not in the genes or in the heads of isolated individuals but in new reactions to concrete social conditions and a reality that is open to exploration.

PRACTICAL INTELLIGENCE

It would be wrong, however, to limit creative problem solving to those who have achieved eminence. Creative problem solving is one of the most commonplace activities and it occurs routinely whenever people confront ill-defined problems with no immediate solution. Sylvia Scribner (1986) defines *practical thinking* as "mind in action" and points out that it is always embedded within larger purposeful activities. These may involve mental tasks, such as choosing a particular product at a supermarket, or manual activities, such as repairing a car. Practical thinking differs from more academic tasks in which thinking is an end in itself. It is purposeful and goal directed, frequently involving an extensive scrutiny of the environment and a search for various alternatives. The ability to think practically varies enormously and many otherwise gifted thinkers seem to be ill-equipped to deal with practical problems in everyday life.

Scribner (1984, 1986) uses several examples of practical thinking by drawing from her research on people in a milk-processing plant. Novice delivery truck drivers, for example, calculate prices by multiplying the unit price by the number of units. Seventeen quarts of skim milk at 68 cents a quart, for example, is calculated by multiplying 17 by $.68. Experienced drivers, on the other hand, develop the capacity to price orders by using a combination of case and unit prices. If a case holds 16 quarts, then 17 quarts equals a case (at $10.88) plus one quart (at $.68). Thirty-one quarts equals two cases (i.e., $21.76) minus one quart. Over time, novices catch on to the simplicity of using case and unit prices and begin to use them as well.

Inventory workers who count the number of products in the warehouse provide a second example. Counting can be done literally—one case at a time—but inventory workers soon learn shortcuts, such as using multiplication to calculate the number of cases in a three-dimensional array. When a large array is not a perfect rectangle, the number of cases in a rectangle is still calculated and cases are added or subtracted to get the total. In both cases, experienced workers redefine tasks to make them as easy as possible. Skilled practical thinking is marked by flexibility, and skilled workers develop a variety of shortcuts to standard operating procedures. Knowledge and skills tend to be highly specific and derived from experience within a particular domain.

What is true of dairy workers is true in other professions as well. Wagner and Sternberg (1986) point out that the typical correlation between IQ and occupational performance is around .2, which accounts for only 4% of the variance. Some argue that this estimate is too low because of the unreliability of the measures and the restricted range of IQs within a particular occupation. Others suggest that it overestimates the actual relationship because other factors, such as social class and education, are eliminated. Wagner and Sternberg (1986) argue that much real world performance is based on *tacit knowledge,* which is not usually expressed or taught. They developed a measure of tacit knowledge and found that it predicted job success but that the correlation between it and IQ was not reliably different than zero.

Numerous examples of highly developed practical skills in nonindustrial cultures are described by Berry and Irvine (1986). Motte (1979), for example, has found that the Biaka Pygmies of the Central African Republic know over 300 medicinal uses for local plants. The Inuits of northern Canada can make remarkably accurate three-dimensional maps, which include elevations and depressions and cover areas up to 250,000 square miles (Bagrow, 1948; Carpenter, 1955). The exceptional navigational skills of the Puluwat people, who use the sun, stars, wave, and wind patterns, among other things, have been described by Gladwin (1970). In each case, native peoples have developed highly particular exceptional skills to meet the demands of their environment. In fact, one of the difficulties encountered in this type of research is that researchers are prone to *underestimate* the depth and breadth of native people's intelligence because they lack sufficient background knowledge to appreciate it.

A more general model of practical intelligence has been developed by Baron (1981), who draws from John Dewey's (1911/1933) description of *reflective thinking.* Reflective thinking involves five different phases—(1) problem recognition, (2) generation of alternatives, (3) reasoning, (4) revision, and (5) evaluation—each with its own set of pitfalls that may limit the quality of the final solution. People probably differ most profoundly in their ability to detect problems. Reflective thinking requires that people recognize problems and suspend judgment until they find an effective solution. Most behaviors occur in a rather "thoughtless" manner by following habitual courses of action.

People also differ in their tendency to *generate alternatives* (Phase 2). To many people, doubt or uncertainty is disagreeable, and they want to end it as soon as possible. The inability to generate hypotheses or consider alternatives is one of the things that separates creative and noncreative solutions. An example can be seen in the technique known as *brainstorming* (Osborn, 1957). Brainstorming is a procedure designed to facilitate creative problem solving in groups. It requires that the group members uncritically generate ideas. Participants are encouraged to avoid editing their own ideas and to say simply whatever comes to mind. Wild, funny, and bizarre ideas are encouraged, as is building on the ideas of others. Suggestions are either tape-recorded or written down by a group secretary so that a large number of ideas can be generated. The group is encouraged to continue to generate ideas until members feel that all options have been exhausted. Once this happens, the group then critically examines the ideas developed in the previous phase. They reexamine the ideas, reworking them when necessary, and select the one that seems most appropriate.

The two stages of brainstorming correspond to the processes involved in divergent and convergent thinking. The reason many people are unable to come up with creative solutions to complex problems is that they do not spend enough time and effort during the initial phase. They settle for routine, habitual solutions that have worked well in the past, or they generate only a limited number of new solutions and take the first one that comes to mind. Many people cannot break away from functional fixedness and avoid bizarre and seemingly inappropriate responses because they do not want to seem odd or unconventional.

During Phase 3—*reasoning*—possibilities are evaluated by examining the evidence for and against each. A common problem during this phase is a "confirmation bias," in which people seek information that supports their initial position and ignore evidence supporting other alternatives. This tendency tends to reconfirm the "rightness" of their original choice so that people who are less open-minded are also the ones who are most sure of the correctness of their decisions. Having never considered alternatives, or having considered them only in passing, the evidence seems very one-sided.

During Phase 4—*revision*—the possibilities are revised on the basis of the evidence. People differ considerably in their willingness to revise their decisions and are generally quite insensitive to counterevidence. Some people select the first alternative which comes to mind, whereas others get stuck in a phase of indecision—carefully weighing each alternative and vacillating from one choice to another.

During the fifth phase—*evaluation*—people must decide whether to make a decision or go back to Phases 2 or 3. They must ask themselves whether further work is likely to pay off. People differ in terms of being too impulsive or indecisive. Impulsive people make decisions quickly before considering all the evidence, whereas indecisive people move back and forth between the various phases unable to make a decision. The term "phases" rather than "stages" is used because the timing is extremely variable and people move back and forth across the various phases during the course of a decision.

Dewey's (1911/1933) theory of reflective thinking can serve as an alternative to standard models of intelligence because it stresses *dispositions* rather than abilities. Effective thinking is based on the tendency to recognize problems, consider alternatives, gather and consider evidence in an unbiased way, and use this evidence when making a decision. People who perform well on traditional IQ tests may be abysmal at making practical decisions and this is why there is little or no relationship between academic performance and success on the job, in social relationships, or in marriage (see Epstein & Meier, 1989).

The focus on dispositions rather than abilities, makes Baron (1981) optimistic that effective thinking can be taught by instructing people to generate more possibilities before evaluating them, weighing counterevidence more heavily, and setting higher standards before making a decision. Impulsive people can learn to slow down. Because these processes are more or less under voluntary control, people can be shown where their previous tendencies have led to bad decisions and taught to overcome their weaknesses and correct their mistakes. They can, for example, use a procedure similar to brainstorming to generate ideas without evaluation and then consider each one before making a decision.

A related question is, *Can creativity be taught?* Aspects of creativity can be taught. People can be taught to keep an open mind, consider more alternatives, weigh the pros and cons and so on, but as we have seen, creativity is based on both conscious and unconscious processes. Many of the unconscious processes, such as chunking, hierarchical organization, speed of information processing, and recall, are based on previous knowledge and experience. Improvements here are based on studying harder and learning more. Creative people also differ from less creative ones by a love of learning, a sense of mission, and a strong desire to excel. It is unlikely that any of these can be directly taught or even appreciably modified through instruction. To be truly creative, a person must follow all the conventional clichés—such as keep an open mind or do your own thing—but this knowledge must be coupled with a drive and persistence not easily taught but without which creativity is impossible.

CHAPTER 11

Formal Thinking

This chapter begins with a discussion of the nature and development of mathematics and logic. Each section includes a brief description of the history, psychological development, and cross-cultural differences. The chapter ends with a review of the recent research designed to teach metacognitive skills to children and the interrelationships among motivation, metacognition, and culture.

As mentioned previously, formal thought differs from informal thought in that formal problems are usually well defined and contain all the information necessary to reach a solution. The distinction between these two types of thought is important because, while informal thought is usually associated with practical problem solving in everyday life, descriptions of formal thought are often used to support the claim that some forms of thinking, such as logic and mathematics, are unconditionally true and do not depend on direct experience. Logical propositions, such as,

If A is greater than B
and B is greater than C,
then A is greater than C,

are always true and it makes no difference whether A, B, and C are apples, chimpanzees, or jet aircraft or whether the comparison is based on weight, age, or any other unidimensional quality. The same is true for mathematical statements, such as $2 + 2 = 4$.

The truth of such statements has led many people to assume that there is a realm of knowledge outside experience not subject to empirical verification. Statements such as the above are assumed to be true even if they are not recognized and, indeed, even if there are no conscious minds to grasp them. The failure to grasp the truth of such statements is seen as a problem within the observer, not a feature of the statement itself. The study of logic and mathematics is the study of correct forms of thinking that are independent of any specific subject matter or even any particular person.

I tend to adopt the position taken by the American pragmatists and, later, by Wittgenstein in which truth and falseness pertain to statements and ideas rather than to empirical facts. Empirical objects and events are not true or false. They simply are. Many of the most elementary aspects of mathematical knowledge, such as numbers, counting, and addition, are discovered anew by each generation by abstracting essential features from direct experience. Each child, for example, discovers that adding one more item to two previous items always creates three, and it does not matter whether the items are buttons, apples, or bananas. But the bulk of contemporary logic and mathematics is a *human invention,* which has evolved gradually over several thousand years. The special status of logic and mathematics is based on the fact that they are tautologies, in which the premises and conclusions or the two sides of the equation are merely different ways of saying the same thing. This can be seen most clearly in mathematics where the two sides of the equation are separated by an equal sign, but it is equally true of logic. In fact, contemporary symbolic logic uses signs and equations similar to those used in algebra and has come to resemble mathematics. This does not mean that the conclusions are immediately and intuitively obvious. By combining and recombining propositions and statements, it is possible to draw truly novel insights. But the ability to reach novel conclusions rests on the assumption that each step in the sequence, like its predecessor, is also a tautology. Over the course of time, people have developed a complex system of logic and mathematics, capable of dealing with a wide variety of practical and intellectual problems, and we are the heirs of these inventions.

But the development of mathematical and logical thought can also be studied from a psychological perspective. Not everyone learns to think logically, and many people fail to develop even rudimentary mathematical skills. These failures are based, at least in part, on the way mathematics and logic are taught in school and certain stereotypic attitudes toward these subjects that make it possible or even desirable to remain ignorant. An adequate understanding of these failures must start with the nature of logic and mathematics—how they have developed over the course of time.

MATHEMATICS

The current widespread use of Arabic numbers, multiplication, division, and so on obscures the fact that mathematics has had a long history, culminating in a body of knowledge so extensive that contemporary schoolchildren can solve problems that were beyond the ability of the most gifted mathematicians in previous times (see Boyer, 1985, for a review). One of the earliest examples of using numbers is a bone of a wolf found in Czechoslovakia that was cut with 55 notches arranged in groups of 5, forming two series of 25 and 30 each. The bone is 30,000 years old and it antedates both writing and civilization, in the usual sense of these words.

Before 4000 B.C., the Egyptians had developed calendars to predict the annual flooding of the Nile, and they later developed surveying procedures for the purpose of reimbursing those who lost land during these floods. If someone lost land, a surveyor was sent to assess the damage and the taxes were reduced accordingly. This laid the foundation for what subsequently became known as "geometry," which stems from the Greek words *geo,* which means earth, and *metron,* to measure. The ancient Egyptians also carried out some impressive bookkeeping feats. A museum at Oxford has a royal mace more than 5,000 years old that records 120,000 prisoners and 1,422,000 captive goats.

In spite of these accomplishments, much Egyptian mathematics had a rough-and-ready quality with little distinction between accurate and rough estimates. In one surveying deed, for example, the area of a piece of land was calculated by adding the two opposing sides. Multiplication was based on doubling—that is, adding a number to itself—and then doubling again if necessary. Procedures of this sort led to only rough approximations rather than accurate sums based on precise formulas. The math developed in Babylon or Mesopotamia was similar in that it was usually applied to practical problems, such as bookkeeping or the inheritance of land. As with Egypt, there was little distinction between exact and approximate measures and little explicit discussion of formal rules by which precise calculations could be made. The lack of such statements has led some to conclude that Egypt and Babylon failed to develop true mathematics, in spite of their high levels of technical proficiency.

The *distinction between pure and applied mathematics* was stressed by the Greeks. Pythagoras (ca. 580–500 B.C.) raised mathematics to the status of a religion and created a secret society with the motto "All Is Number." Mathematics became a branch of philosophy, and discussions became so abstract that little consideration was given to practical application. The Greek distaste for applied mathematics reflected to a certain extent their distaste for practical affairs in general, which were usually left to slaves. The Greeks made a sharp distinction between *arithmetic,* based on the theory of numbers, and *logistics,* which were routine mechanical computations. Logistics was seen as the domain of merchants and military men, who must understand numbers in order to arrange their troops. Philosophers, in contrast, were concerned primarily with pure mathematics. The institution of slavery fostered a separation between theory and practice, leading to a neglect of experimentation that hindered the development of Greek science.

The Greeks drew from Egypt and Babylon, but their mathematics was radically different. Mathematics before Greece was mainly a collection of empirical observations to deal with practical, everyday problems. For the Egyptians, a straight line was a piece of stretched string and a rectangle was a bound piece of land. The Greeks made mathematics abstract. As Kline (1953) expressed it, "The Greeks eliminated the physical substance from mathematical concepts and left mere husks. They removed the Cheshire cat and lift the grin" (p. 31). Greek geometry dealt with points, lines, planes, and circles that were seen as the "essence" of objects in the physical world.

The Greeks also insisted that all mathematical conclusions be derived by *deductive reasoning*. Euclid (350–300 B.C.) started with 10 axioms, such as the shortest distance between two points is a straight line, and derived hundreds of theorems. When these were later applied to objects in the real world, they seemed to fit perfectly. All triangles, for example, appeared to have interior angles that summed to 180 degrees, and the area of various figures could be calculated precisely. Although these beliefs were later challenged by non-Euclidian geometry, the development of Euclidian geometry demonstrated the power of pure reasoning and helped foster a rational spirit within the Western world.

Despite the unprecedented gains in geometry, trigonometry, and astronomy, Greek mathematics had some serious weaknesses. The most obvious was the use of the Greek alphabet to represent numbers. The Greeks used the first nine letters to represent the numbers 1 through 9, the second nine letters to express multiples of 10 (i.e., 10—90) and the third nine letters for multiples of 100. Numbers greater than 1,000 were represented by repeating the sequence with a stroke or an accent. The number 625, for example, was represented by the letters XKE and any number less than 10,000 was easily written with only four characters. Such a system makes notation easy but calculations difficult. The Romans, who had little interest in pure mathematics, reverted to an even cruder system based on Roman numerals.

At about the same time that the Greeks were developing their math, tremendous progress was being made in China and India. The Chinese developed a system of "rod numbers" in which the numbers 1 through 5 were represented by one to five strokes, 6 through 9 used a perpendicular line and one to four strokes, and a blank space was used to represent zero. By alternating horizontal and vertical patterns, any number could be represented, and calculations could be carried out easily by administrators who carried bamboo, ivory, or iron rods. The use of a base-ten system meant that decimals were easily incorporated and the Chinese had little problems with negative numbers, since they used a black set of rods for positive integers and a red set for negative ones. Meanwhile, the Hindus in India made several advances. They incorporated the concept of zero and used a notational system for numbers 1 through 9 that was the forerunner of our own "Arabic" numerals.

Some of the greatest advances in mathematics came from the Arabs during the Middle Ages. They combined the Greek concepts of geometry with the Hindu number system, introduced algebra, and preserved many of the ancient texts that would have otherwise been destroyed by church authorities. The Arabs had a great respect for Greek culture, but they also recognized the utility of the nine Hindu digits and kept mathematics alive during the Dark Ages of Western Europe.

Despite the superiority of the Hindu–Arabic number system, the transition from Roman numerals in Europe was surprisingly slow and very little real progress was made during the Middle Ages. During the Renaissance, mathematics was more widely used in bookkeeping, construction, and art. There was also a revival of interest in the classical Greek works. By 1575, Western Europe had recovered most of the major classical works on mathematics, Arabic algebra had been mastered and improved, and Western Europe was on the verge of several breakthroughs.

All the discoveries prior to the 1600s represent only a tiny fraction of what has become contemporary mathematics. Analytic geometry and calculus were invented during the seventeenth century, probability theory somewhat later. The nineteenth century saw the rise of non-Euclidian geometry and has sometimes been called the "Golden Age" of mathematics because the number of discoveries far exceeded all those occurring in previous times. The present century has seen further developments in correlational and statistical procedures, which have been increasingly linked to the development of computers.

Even a brief look at the history of mathematics should dispel the notion that mathematics is a static entity. The "timeless" truths of mathematics are not timeless in the sense that they have always been with us or in the sense that there is an ethereal realm in which formulas and equations are waiting to be plucked like ripe tomatoes from the vine. Many of the discoveries have occurred because of a need to deal with practical problems in everyday life. Others are the result of intellectual challenges left unanswered by previous mathematicians. The end result is a substantial body of mathematical knowledge whose content should be familiar to most educated people.

The practical nature of mathematics can be seen most clearly in contemporary studies of *nonindustrial peoples.* Almost all cultures have had to deal with practical problems, such as barter and exchange, so it is not surprising that virtually all cultures have developed methods of counting and measuring quantity. Most rely on fingers and toes as aids and, therefore, take the form of a base-ten system. Remnants of this can be seen in our own language where we use the term *digits* to refer to both numbers and fingers and toes. The degree of mathematical sophistication, however, varies from culture to culture according to the practical needs of a given society.

Gay and Cole (1967), for example, studied the Kpelle people of central Liberia and found that they were remarkably accurate at estimating the number of cups of rice in a bowl. Because rice is a staple and frequently exchanged, judgments of this kind occur often and, presumably, improve with practice. Rice was measured in one of two ways. The basic unit of measurement was a tin can holding about two cups. The bottom was pushed out and rounded when the rice was bought and flattened when it was sold. The difference in volume was one source of the rice traders' profits. It was also sold by bags that contained nearly a hundred cups. Another unit of measurement was the Coca-Cola bottle cap, which was used for selling snuff.

The Kpelle also had ways of measuring length, time, and money but not weight, speed, or temperature. Length was measured by the distance between the fingertips when the arms are outstretched or between the fingertips and the nose. Handspans were used for smaller items. Time was measured in terms of literal days—sunrise to sunset. Weeks were the units between market days; months were based on the cycles of the moon; and years were the time between planting seasons. What is striking about all these units for measurement is that they were not standardized or related to each other in any systematic way. Time, for example, was seldom measured in units greater than two. Some people did not know the number of months in a year, and most people did not know their own age. Measurement was used only where it was most needed, and approximations were used in other cases.

Other cultures have developed similar strategies. The Oksapmin of New Guinea count by using a system of 27 numbers, each based on a particular part of the body (Saxe, 1981). The fingers on the right hand represent 1 through 5. Right wrist, forearm, and elbow are 6, 7, and 8. They then count up the arm, over the shoulder, around the head and over to the left arm. Other cultures use kola nuts or cowry shells. West Africans, who used cowry shells when trading with Portuguese merchants in the seventeenth century, could calculate sums running into the tens of thousands.

What characterizes mathematics in all these cultures is that it is based on practical needs. People in nonindustrial cultures develop mathematical skills to deal with practical problems, and they fail to develop skills when they are not necessary. Ginsberg (1977, 1983) suggests that certain elementary forms of mathematical knowledge may be universal and cut across both social classes and cultures. Children in all cultures must simultaneously internalize a body of abstract knowledge based on numerical concepts and their relationships and develop strategies that allow them to deal effectively with the problems in their culture. Some of these skills occur quite early. Before 6 months of age, infants can distinguish arrays with one to four items (see Klein & Starkey, 1987). By 6 months, they notice the similarity between audio and visual displays—looking, for example, at two objects rather than three when two sounds occur. Two-year-olds learn to count to two and, sometimes three, using conventional numbers. In each case, children develop rudimentary mathematical concepts by abstracting essential features from concrete situations, but this abstract knowledge is eventually supplemented by a number of additional skills to deal with more difficult problems.

Hitch, Cundick, Haughey, Pugh, and Wright (1987) suggest that children and adults judge numbers in two ways. Arrays less than five are seen at a glance and this helps to explain why illustrations of small sums, such as $2 + 2 = 4$, are immediately and intuitively obvious. This phenomenon has been known for some time, and there has been a steady trickle of research on it since the 1940s. Kaufman, Lord, Reese, and Volkman (1949), for example, suggested that the ability to identify small numbers is so fast and accurate it must involve a special process, which they called "subitizing." Subitizing seems to be based on the shape of a configuration, and it is used to judge numbers up to four or five. Klahr and Wallace (1976) have found evidence of subitizing in children as young as three or four.

Larger arrays are counted subvocally. Preventing counting interferes with the judgment of arrays greater than four. Hitch and his associates compared children and adults and found that 6-year-old children counted faster aloud than subvocally, but these differences gradually diminished and ultimately disappeared. These results support the idea that internalized speech requires effort but becomes more and more automatic over time. The speed of both overt and covert counting also increases throughout childhood.

School places additional demands on children and forces them to develop additional strategies. Children in industrial countries are expected to learn a much more complex system of mathematics, which has little or no immediate application. Siegler (1987) found that children use a variety of strategies when adding. These include counting fingers, counting aloud, putting up fingers for each number and using direct retrieval from items previously learned and stored in memory. Most children use a number of different strategies, even when dealing with the same problem, but they do show preferences for one over the others. Siegler points out that the ideal strategy should be both fast and accurate, but these two criteria are somewhat incompatible. Direct retrieval is both the fastest strategy *and the least accurate* and it therefore tends to be used primarily on simple or well-learned problems.

Resnick (1983) traces the development of children's mathematical knowledge as they move from simple counting strategies to more complex forms of understanding. Preschool children seem to treat numbers as a string of items in which each one in the sequence is larger than the previous one but smaller than the next. This primitive understanding allows preschool children to carry out certain basic operations, such as counting or comparing numbers in terms of more or less. Schoolchildren develop a more sophisticated understanding of numbers based on part–whole relationships, in which small numbers are seen as components of larger ones. These are usually based on sets of three, such as 3, 4, and 7, which can be used to solve a variety of problems, such as $3 + 4 = 7$, $4 + 3 = 7$, $7 - 4 = 3$ and so on. Well-learned associations of this sort are often used informally to solve more difficult problems. Five plus four, for example, may be solved by adding $4 + 4$ and then one more. These shortcuts are not taught in school but evolve through extensive practice and serve as proof that children are developing rather sophisticated forms of procedural knowledge that they can use flexibly to deal with a variety of problems. What is more, children seem to develop and use such strategies without being able to understand and describe what they are doing. Later, such shortcuts are extended to include multiples of 10. A problem, such as 32 + 17, for example, is solved by adding 10 to 32 and then seven more. Children, thus, become able to deal effectively with multiple digits involving units of tens, hundreds, and so on. Although these procedures are invented informally by schoolchildren in our culture based on extended exposure to mathematical problems of a similar nature, Saxe and Posner (1983) provide evidence that similar strategies occur among illiterate Africans, as well. As Resnick (1983) points out, one consequence of this bottom-up model is that it forces investigators to conceptualize the development of mathematical ability as a large number of relatively small changes in mathematical knowledge, thus providing a *microstage* theory for numerical understanding that includes both procedural knowledge and conceptual understanding.

Whereas addition and subtraction are learned primarily through concrete examples, in which children see and manipulate objects, multiplication and division are learned primarily through drill. As Campbell (1987) points out, most adults take for granted their ability to multiply figures such as 6 × 9 and forget the five or six years of practice required to master these fundamental skills. Campbell and Graham (1985) studied adults and found that the rate of error for some combinations, such as 4 × 8 or 6 × 9, approached 30%. Ninety-three percent of these errors were based on correct answers to other problems, such as 6 × 9 = 56 or 63. They also found a strong positive correlation between errors and reaction time. People tend to take a longer time answering questions in which errors are common, and this is apparently due to competing responses from false candidates even when they are not selected. This pattern suggests that people learn a variety of false associations when they learn their multiplication tables, and these are later weakened through practice. Campbell (1987) and Graham (1987) also suggest that people's ability to multiply small numbers better than large ones is not due entirely to size. Multiplication of small numbers occurs earlier and, therefore, produces fewer competing responses. These problems are also practiced more often and occur more frequently in textbooks. Graham (1987) found that children who were taught multiplication in a mixed order showed a smaller correlation between error rate and size.

Finally, Greer (1987) has found a close connection between the difficulty of mathematical problems and the way they are worded, which erases to a certain extent the traditional distinction between math and verbal ability. Using the following examples provided by Vernaud (1982), he points out that the same problem—that is 4 + 7—can vary greatly in the degree of difficulty:

1. There are 4 boys and 7 girls around the table. How many children are there altogether?

2. John has just spent 4 francs. He now has 7 francs in his pocket. How much did he have before?
3. Robert played two games of marbles. On the first game, he lost 4 marbles. He played the second game. Altogether, he now has won 7 marbles. What happened in the second game?

This suggests that many of the difficulties that people experience in school and on standardized tests that measure mathematical ability are not caused by deficiencies in mathematics per se. They are often due to an inability to figure out what is required. People who cannot understand the problem cannot answer it, even when they have the necessary skills.

Because the ability to solve problems depends on the ability to understand what is being asked, children develop strategies to help them deal with problems in which the instructions are not clear. These include (1) finding the numbers and adding, (2) guessing the operation, (3) calculating all the alternatives and selecting the most reasonable one, or (4) looking for key words, such as "more" or "less," which signal the most likely operation. People can also infer the operation from the numbers. If the numbers are 78 and 54, for example, they can try addition or subtraction. If they are 78 and 3, division is a more likely alternative. The problem with these strategies is that they are undeservedly successful. They work most of the time, so they become part of children's arsenal of tricks to deal with problems.

Children also develop a number of misconceptions about arithmetic based on implicit theories about operations. These include the assumption that multiplication makes bigger and division makes smaller. These are based on a tendency to see multiplication as repeated addition—to see 3×4, for example, as $4 + 4 + 4$. Although this is true of whole numbers, it is not true of fractions. As a result, children often experience difficulty when one of the numbers is less than one. Greer points out that these informal strategies should be recognized. Teachers should be more aware of children's misconceptions about mathematics and correct them by using specific examples.

In short, research on the development of mathematical skills shows that they are strongly affected by processes not typically associated with mathematics. These include the speed of recognition, strategies used in addition and subtraction, suppression of false associations during multiplication and division, and the ability to understand what is being asked. All of these are rather rudimentary skills, but they are the foundation on which more advanced skills are based.

As a result, people vary widely in mathematical ability. One of the most noticeable differences occurs between males and females. Girls and boys in early elementary school receive similar grades and perform equally well on standardized tests for measuring mathematical ability, but differences between the sexes increase throughout elementary school. Benbow and Stanley (1980, 1987) examined the scores of nearly 10,000 mathematically talented children in the seventh and eight grades and found that boys scored 30 points higher on the mathematical component of the Scholastic Aptitude Test, which is designed to screen candidates for entrance into universities. Differences between boys and girls were particularly pronounced at the upper end of the scale. The ratio of boys to girls scoring 500 or more was 2 to 1. Among those who scored 600 or more, the ratio was 4 to 1, and for those extremely rare junior high school students who scored over 700, boys outnumbered girls by a ratio of 13 to 1. Because boys and girls of this age have a similar background in mathematics, Benbow and Stanley attributed these differences to differences in innate ability.

Others disagree. Eccles and Jacobs (1987) point to a study by Leinhardt, Seewald, and Engle (1979) that suggests elementary teachers spend more time teaching mathematics to boys. As a result, boys may receive as much as 36 more hours of instruction by the time they reach the seventh grade. Boys are also more likely to have informal contact with mathematics through toys, books, and games that contain mathematically challenging activities (Astin, 1974). Differences between boys and girls are also strongly influenced by *parental expectations* and the value they place on mathematics. Despite similar levels of performance in elementary school, mothers and, to a lesser extent, fathers tend to rate the mathematical ability of their daughters lower than parents with sons and come to expect that their daughters will experience more difficulty with math in the future. Parsons, Adler, and Kaczals (1982) have also found that parents believe their daughters must work harder to do well and ranked math as less important than other subjects, such as English and history. These beliefs cause young girls to underestimate their own level of ability, experience math anxiety, and systematically devaluate mathematics.

As a result, girls tend to avoid mathematics in high school. Sell (1978), for example, studied the mathematical background of first-year students at the University of California at Berkeley and found that 57% of the incoming men had four years of high school mathematics, but only 8% of women had a similar level of training. This meant that 92% of first-year women could not enroll in 15 of the 20 majors that required a strong background in math, such as engineering, physics, or computer science. The presence or absence of high school mathematics acted as a "critical filter" that reduced the number of potential majors for women and, as a consequence, their range of occupational choice.

One of the reasons people fail to develop math skills is that lack of ability is socially acceptable and has in some circles achieved the status of a social grace. The division of knowledge into science and humanities means that many people specializing in humanities not only fail to take math seriously but take pride in their inability. As Papert (1980), Dweck (1986), and others have pointed out, difficulty with mathematics in school is often the first step in an evasive process in which people learn to see themselves as bundles of aptitudes and ineptitude and begin to consider themselves unmathematical, not artistic, unmusical, or even stupid.

Papert (1980) recommends using computers to teach math and higher order skills that improve learning in general. Today, computers are used mainly to present problems and provide feedback about whether the answers are correct or not. Computers serve as a substitute for traditional instruction and allow children to learn at their own pace. But Papert has shown that even young children can learn to program computers to carry out basic tasks, such as drawing or composing music. This allows them to learn basic mathematical and scientific concepts within a context of playing games or solving puzzles. Computers make mathematics concrete, and their "holding power" makes learning fun.

The nature of computer programming is also such that mistakes are common. A child trying to draw an equilateral triangle, for example, might instruct the computer to draw a line 3 inches long, turn 60 degrees, draw another 3-inch line, turn 60 degrees, and then draw another one. If they did, the child would discover that the computer had drawn one-half of a six-sided figure rather than a triangle. To draw a triangle, they would need to instruct the computer to turn 120 degrees each time. By making mistakes and correcting them, children learn that it is not necessary to get things right the first time. What is more important is the ability to break the problems down into "bit size" steps that can be dealt with separately. This new knowledge is a source of power, and it can give children who previously had difficulty with mathematics a liberating sense of being able to do things which were previously considered "too hard."

LOGIC

As Bertrand Russell (1917) has pointed out, formal logic began when Aristotle (384–322 B.C.) introduced the notion of syllogisms and it came to form one of the two major areas of academic concentration (along with theology) during the Middle Ages. Although many of the most gifted thinkers of the Middle Ages were devoted to the study of logic, very little real progress was made until the middle of the nineteenth century. According to Russell (who tended to exaggerate), more progress was made during each decade since 1850 than in all the preceding periods. It is now known that both the Stoics in ancient Greece and the Schoolmen during the Middle Ages made significant contributions that were simply forgotten or overlooked by later thinkers. Syllogisms are now only a tiny part of modern logic, which now employs far more symbols and has become more and more like mathematics.

Logic represents correct forms of thinking, not psychological processes. It is composed of verbal or symbolic descriptions of the way people *should* think if they are to arrive at correct conclusions. This can be seen most clearly in the case of syllogisms. *Syllogisms* are forms of discourse in which, when certain things are stated, others follow necessarily. They contain three statements and three terms and generally move from general to particular. A common example is

All men are mortal.
Socrates is a man.
Therefore, Socrates is mortal.

Syllogisms allow people to draw specific conclusions once certain general conditions are known.

Syllogisms were considered the principal form, perhaps the only form, of logic until the nineteenth century, but modern logic differs from Aristotelian logic in several ways. It is much more general, deals with a variety of different forms of reasoning not previously considered, and, as mentioned earlier, employs far more symbols and is more like mathematics. The foundation for modern *symbolic logic* began when George Boole (1815–1864) noted that the letters and notation used in algebra need not be limited to numbers. If x, for example, stands for people who live in the United States and y stands for those with blue eyes, then $x \times y$ (or more simply xy) represents the number of people living in the United States with blue eyes. The statement $x + y$ represents the number of people who live in the United States *or* have blue eyes, and so on. Using these and similar notations, it became possible to express logical relationships in symbolic form.

Further progress was made when additional symbols were introduced. These include universal quantifiers (x) and existential quantifiers $(\exists x)$, which correspond more or less to "every" and "some." The negative sign (or $\sim$) for negation, $\supseteq$ for "If . . . then," and conjunctions, such as & for "and" and *v* for "or" were also introduced. Traditional sentences can now be translated into symbols; parentheses can be used to group items; and it is now possible to produce long sequences, such as

$$(x)(Px \supset Mx) \supset -(\exists x)(Px \;\&\; -Mx)$$

that resemble mathematical equations. The left side of the preceding can be used to express the statement "All people are mortal," and it is equivalent to saying, "It is false that there are some people who are not mortal." It is simply a different way of expressing the relationship between men and mortality, but it illustrates how different modern symbolic logic is from the traditional logic introduced by Aristotle. The close relationship between logic and mathematics led many of the leading thinkers at the turn of the century to claim that mathematics is a form of logic—a form of thinking by which the user draws valid conclusions from arbitrary premises.

As Barker (1965) points out, although logicians normally focus on correct forms of thought because these are the most theoretically interesting, from a practical point of view it is equally important to recognize *fallacies.* These come in three major forms. First are *inconsistencies* in which people use premises that are not consistent or are mutually contradictory. Then there is the *petitio principii* or "begging the question" in which the premises are more dubious or difficult to accept than the conclusion. Finally, there are various forms of *non sequiturs* or pure fallacies in which the conclusion does not follow from the premises. One of the more common forms of non sequiturs is ambiguity in which the meaning of a term is altered. An example is

Time is money.
Time is measured in seconds.
Therefore money is measured in seconds.

The term *time* is used here both figuratively and literally. Because the meaning of a word can change radically within a proposition, proper conclusions cannot be drawn by comparing individual terms in isolation. The ability to think logically and spot fallacies depends on practice, knowledge, and a much more general understanding of language itself.

There has been considerable debate about the *nature of logic.* At one time, it was widely believed that logic described actual mental processes—how we think and, perhaps, how we must think (see Haack, 1978). Immanual Kant (1724–1804), for example, felt that the principles referred to in Aristotelian logic described the "laws of thought," without which thinking would be impossible. Others, such as F.L.G. Frege (1848–1925), have argued that logic is an independent entity and has nothing to do with thought. Frege felt that logic is objective and public, whereas thoughts are private and subjective. In this view, mental processes are not the subject matter of logic, and logic is not a branch of psychology. The subject matter of logic includes terms, sentences, and arguments embodied within language itself. This distinction between objective and public and internal and private is more difficult to accept, given Vygotsky's (1934/1962) thesis that thoughts are based on internalized speech. Perhaps the most common position is that logic represents *correct forms* of thinking—the way we must think if we are to reach the proper conclusions.

One of the strongest attacks on traditional logic was launched by John Stuart Mill (1806–1873). Mill (1843/1949) argued that nothing can be proved by syllogisms that was not already known and that every syllogism involves the *petitio principii* in which the premises are more difficult to accept than the conclusion. Take the familiar example:

All men are mortal.

Socrates is a man.

Therefore, Socrates is mortal.

Mill argued that, to know that all men are mortal, one must know that each and every man, *including Socrates,* is mortal. Socrates' mortality is guaranteed by the more general statement in the premise and, therefore, the conclusion tells us only what we already knew. Mill pointed out that syllogisms are basically circular and that deductive arguments cannot be used to prove anything because it is necessary to infer general truths before deriving specific conclusions. Such statements are not worthless, since they can serve as a check on the accuracy of our thinking, but they do not provide us with a means of deriving *necessary truths,* which were not already known.

Hillary Putnam (1971) has pointed out that logic, like every branch of knowledge, undergoes changes during the course of its development but the changes are somewhat different from those in other areas. Logicians in different centuries have had different ideas about the scope of logic, the procedures used, and the relationship between logic and thought. Propositions once established, nevertheless, remain true. The preceding example about Socrates is as true today as when it was first introduced, but the explanation is somewhat different. Today it is more common to explain these relationships in terms of set theory—if men are a subclass of those things that are mortal and Socrates is a subclass of men, then Socrates is also a subclass of those things that are mortal. Although ancient and modern thinkers would agree about the truth of the statement, the interpretation is radically different, and it is a far cry from what traditional logicians thought they were saying when they talked about the necessary laws of thought. In other words, logic changes, not because previously true propositions are rendered false, but because the style and interpretation vary considerably and the scope of logic has been expanded to include areas not considered previously. What is even more disturbing is that there is still considerable controversy over what the "correct" interpretation is.

If we now ask, *Do people think logically?* the answer depends on how we define logic. If logic is defined as a system of formal rules that are explicitly stated and used to guide thought, then it is apparent that few people reach this level of intellectual development. Of more interest is the question whether people *implicitly* use these rules in drawing conclusions. The distinction is similar to that made previously between people's ability to use grammar and the more formal description of grammatical rules provided by linguists. Once the question is phrased in this way it becomes possible to study the development of logical operations in children and compare people in different cultures.

One of the earliest attempts to study "cross-cultural" differences in logical thought was carried out by Luria and Vygotsky in the Soviet Union (see

Luria, 1979). As Wertsch (1985) has pointed out, this research is more appropriately called "cross-historical," because Luria and Vygotsky were concerned with changes in logical thinking as regions of the country were developed. After the revolution, the old class structure was eliminated, schools were set up, and new forms of technology and collective farming were introduced. By taking advantage of the developments in remote regions of the country, Luria and Vygotsky hoped to trace changes in logical thought.

Literate and nonliterate subjects responded very differently to *syllogisms*. When presented with a syllogism, literate subjects solved it in the expected way, but nonliterate subjects often refused to accept the premises as universally valid, see the logical relationship between the premises and the conclusion, and draw *any* conclusion at all. These subjects could make excellent judgments and draw proper conclusions about facts that directly concerned them but they were unable to deal with abstract concepts not based on personal experience.

These studies compared nonliterate peasants without formal education with literate subjects who had been to school, and, therefore, it was not possible to separate the effects of literacy and education. Cole and Scribner (1974; Scribner & Cole, 1981) later carried out a study in which education and literacy were separated. They compared subgroups of Vai subjects in Liberia who learned to read with and without formal education. One group learned an indigenous script and used it to keep personal records and correspond. A second group learned Arabic primarily through rote memory for the purpose of reading and reciting the Koran. A final group learned to read and write English in school. Cole and Scribner found that neither of the first two groups showed a shift toward syllogistic reasoning. These results support the claim that formal education rather than literacy per se is associated with the development of higher order intellectual skills.

The differences between formally educated and noneducated subjects were due (at least in part) to an ability to break away from familiar context and treat propositions abstractly. As Scribner and Cole (1973) point out, "School represents a specialized set of educational experiences which are discontinuous from those encountered in everyday life and . . . it requires and promotes ways of learning and thinking which often run counter to those nurtured in practical daily activity" (p. 553). When presented with logical problems such as syllogisms, those who have not attended school often refuse to accept the assumptions stated in the premises and, therefore, fail to draw the implied conclusion. Such subjects tend to solve individual problems one at a time—treating each one as a new problem—whereas those who have been to school treat them as specific examples of related problems that can be solved by applying more general rules. School also encourages a more detached attitude toward thought itself, which tends to encourage the development of higher order metacognitive skills and, in some cases, a change in attitude toward intelligence itself.

METACOGNITION AND MOTIVATION

It is becoming increasingly clear that there is a close relationship between metacognition and motivation and that both are developed and maintained through interactions with significant others within a cultural context. The role of metamemory has already been discussed in Chapter 9, but metamemory is only one form of metacognition. It also includes processes involved in learning, problem solving, and communication.

The concept of metacognition has changed considerably since it was first introduced during the 1970s. Pressley, Borkowski, and O'Sullivan (1984, 1985) have divided metacognition into a number of components. These include specific strategy knowledge about different strategies, such as repetition, organization, elaboration, and the use of mental images, as well as knowledge about how, when, where, and why to use them. Another component is relational strategy knowledge, which is comparative information about the similarities and differences among strategies. Then there is general strategy knowledge or the recognition that learning can be improved if strategies are used effectively. This type of knowledge is especially important for linking metacognition and motivation because it includes the knowledge that strategies require effort but that this effort can pay off in terms of increased understanding and recall. In addition, there are higher order executive processes, which are used to monitor the acquisition and effectiveness of different procedures. In general, metacognition includes any form of declarative knowledge about a person's own cognitive processes and current level of ability. Because it is a form of declarative knowledge, it can be verbalized and taught to others through explicit instructions. It resembles what Sternberg (1985a) has called higher order executive processes used to plan, monitor, and evaluate ongoing events.

Anne Brown (1975) has suggested that a single metacognitive skill is central—that is, the tendency to develop new strategies—and there appears to be considerable variation among individuals and cultures. Borkowski and Turner (1990) point out that many children are nonstrategic and fail to show a high level of metacognitive knowledge. Many of these differences can be directly linked to differences in motivation and cognitive styles. Borkowski, Peck, Reid, and Kurtz (1983), for example, studied metamemory and strategy transfer in reflective and impulsive children in Grades 1 through 3 and found that reflective children had more metamemory knowledge and were more strategic. Because impulsive children tend to perform poorly on a wide range of tasks, a number of investigators have attempted to train impulsive children to be more reflective by talking to themselves (e.g., Meichenbaum & Goodman, 1971). Rohrkemper and Bershon (1983) suggest that teachers can directly teach children how to use constructive inner speech by telling them what to say to themselves as they engage in tasks and encounter difficulties.

There is also a close link between metacognition and attributional style. As mentioned in Chapter 4, people differ in whether they attribute success and failure to effort or ability (Diener & Dweck, 1978; Dweck, 1975). Those who attribute success to ability and failure to lack of effort tend to be mastery-oriented and more persistent in the face of failure. They tend to regard difficult tasks as a challenge, become more intrinsically motivated, increase effort, and improve their performance over time. They also seem to have an "incremental" view of intelligence and believe that intelligence can be increased by developing new skills. Those who attribute success to effort and failure to lack of ability tend to develop a helpless pattern, in which they become demoralized when faced with potential failure, reduce effort, lose interest, and find it difficult to concentrate. They tend to see intelligence as a stable trait that cannot be improved with practice and believe that there is an inverse relationship between effort and ability—that is, people who are really smart do not need to try very hard and those who do try hard are not really smart. Diener and Dweck (1978) found that mastery-oriented and helpless children differed in the amount of self-monitoring and task-related thoughts. Kurtz and Borkowski (1984) found that children who attribute differences in performance to the degree of effort tend to have more metacognitive knowledge about memory strategies than those who attribute them to uncontrollable factors, such as luck or ability. Mastery-oriented children seem to spontaneously acquire metacognitive knowledge as they confront difficult situations and develop new skills.

Other studies have taught young children metacognitive skills to improve their performance in a particular area. Some of these studies have already been mentioned. Palincsar and Brown (1984), for example, taught poor readers to clarify difficult material, ask questions, summarize what has been read and predict what will happen through a method of reciprocal teaching and found significant improvements in reading scores (see Chapter 1). Paris and Oka (1986) used a similar procedure with approximately a thousand children in the third and fifth grade. They taught children to think about the title and topic before reading, to stop periodically and paraphrase the material, and to skim and reread as a review. Students were also given the opportunity to discuss when strategies should and should not be used. Students in the experimental classrooms improved more than similar controls and had significantly more metacognitive knowledge. They were also more motivated and strategic, and these differences were significant even after the effects of IQ were removed.

There appears to be a *reciprocal relationship* between metacognition and motivation (see Borkowski & Turner, 1990). High levels of achievement motivation and self-esteem and an internal locus of control help determine the choice of tasks, the degree of persistence, and the acquisition of new skills. The development of new strategies increases the awareness that these skills can be used effectively and be transferred to new tasks and situations. This, in turn, increases people's perception of their own ability and gives them confidence that effort can improve performance and master new skills, thus promoting higher levels of achievement motivation, self-esteem, and perceived control. In short, motivation plays a key role in the "spontaneous" development of cognitive strategies and metacognitive knowledge by providing children with the incentive to learn new skills and transfer them to new domains.

Some of the most exciting research in recent years has attempted to increase metacognitive knowledge among young children while observing the effects on motivation and attributional styles. Brown and Reeve (1987), for example, have noted that the method of reciprocal teaching increases not only reading comprehension but facilitates the perception of competence and control. Reid and Borkowski (1987) used a combination of specific strategy training, instruction in executive functioning or self-control, and attribution retraining and found a significant change in metamemory, attributional style, and, to some extent, impulsive behavior. They argue that the inclusion of executive processes and motivational beliefs was essential for producing generalized learning and problem solving and may have set in motion a chain of events that went well beyond the experimental situation.

Similar results have been found by Bandura and Schunk (1981). Children who were very poor in mathematics and showed little interest underwent a program of self-directed learning with proximal subgoals, distal goals, or no goals at all. The material included seven sets of problems with eight different types of subtraction. Those with proximal subgoals were encouraged to complete one set of problems each session, whereas those with distal goals were encouraged to complete all seven sets by the end of the seven sessions. Those who broke the tasks down into a number of challenging subgoals learned quickly, achieved a substantial level of mastery, and developed a sense of personal efficacy and intrinsic motivation. Adopting proximal rather than distal goals seems to have a number of different effects. It increases interest and motivation by allowing people to evaluate their current performance against internal standards; it allows people to more closely monitor and direct their ongoing behavior; and it can increase people's perception of their own ability. This increase in perceived self-efficacy can, in turn, help determine the choice of tasks, the amount of effort, and how long people will persist in the face of potential failure. The tendency to focus on proximal rather than distal goals is somewhat similar to Papert's (1980) suggestion that teaching children to program computers allows them to break complex tasks into a number of more manageable "bit size" steps.

The development of metacognition seems to be essential in the perceived link between cognition and motivation. There is substantial evidence that children can develop and use strategies before they can describe what they are doing (e.g., Sodian, Schneider, & Perlmutter, 1986) and that some skills may not transfer and remain domain specific. Drawing from Hatano (1982),

Brown and Campione (1984) make a distinction between routine and adaptive experts. People who practice skills frequently can become routine experts and perform very well in a particular area. Adaptive experts use strategies more flexibly, modify them to fit particular tasks, and develop new strategies to deal with new problems. Differences between routine and adaptive experts seem to be based primarily on the amount of declarative knowledge. Adaptive experts have more knowledge about how, when, and why particular strategies should be used and a more thorough understanding that they can improve their performance if these skills are employed. Many of the problems associated with transferring skills are based on recognizing that the current problem is similar to those already encountered. They are based on noticing the similarity rather than having the necessary skills. Metacognition increases the efficient use of strategies because it tells people when and where to use specific skills.

CULTURAL DIFFERENCES IN METACOGNITION AND MOTIVATION

There is a growing body of evidence that both metacognition and motivation are shaped by culture. Schneider, Borkowski, Kurtz, and Kerwin (1986) found that German children are more likely to use memory strategies and recall significantly more material than American children of the same age. American children, in contrast, profited more from direct training of rehearsal and organizational strategies and were more likely to attribute differences in ability to the degree of effort. They were twice as likely to attribute differences in performance to effort and this was significantly related to strategy use before and after training in American children but not among Germans.

These characteristics were later traced to differences in the home and in the classroom (Carr, Kurtz, Schneider, Turner, & Borkowski, 1989; Kurtz, Schneider, Carr, Borkowski, & Turner, 1988). German parents reported more direct instruction of strategies, were more likely to check their children's homework, and possessed more games in which strategic thinking was required. Although the limited number of teachers in these studies prevented a meaningful comparison, the trend for teachers was the same. German teachers reported more direct instruction of memory strategies. Differences in attributional styles between German and American children were mirrored by parents and teachers. American parents and teachers were more likely to attribute differences in performance to the degree of effort.

Similar contrasts in attributional styles have been found between American children and those in Iran. Iranian children attribute success to ability more often than those in the United States (Salivi, Maehr, & Gillmore, 1976). In striking contrast, Stevenson, Lee, and Stigler (1986) compared children and mothers in China, Japan, and the United States and found that, although all three groups tended to attribute success and failure to effort most often, American children and their mothers had a greater belief in ability. American mothers were more likely to say that mathematics and reading ability is innate. Americans, therefore, appear to stand midway between Germans and Iranians, on the one hand, and Chinese and Japanese, on the other. These studies suggest that attempts to change children's performance by changing attributional styles (e.g., Dweck, 1975) may be limited because such beliefs are anchored in the broader social context. These beliefs are taught directly through explicit instructions and indirectly through implicit attitudes about the reasons for academic success and failure.

Classrooms can affect motivation and metacognition in a variety of ways. Dunkin (1978–1979), for example, observed elementary school classrooms during the 1970s and found no direct instruction in reading comprehension. Workbook exercises, small group activities, and questions about content were the norm. Duffy, Roehler, Book, Meloth, and Vaurus (1984) compared effective and ineffective teachers and found that effective teachers spent more time teaching strategies and making cognitive processes known. Kurtz, Borkowski, and Deshmuich (1988) compared teachers in Indian and Western schools and found that teachers in India were more inclined to use rote teaching methods and gave little instruction in general learning principles that could foster metacognitive awareness. The performance of children on memory and metamemory tasks reflected these differences in teaching styles.

But classrooms can also facilitate or inhibit the development of cognitive skills indirectly by shaping children's perception of their own ability. Carole Ames (1987) has reviewed the literature in this area and has concluded that when children compete with each other, they tend to compare their performance with each other and make ability attributions for success and failure. Those who do well infer that they have high ability, whereas those who perform poorly come to assume that they are not as smart. Students' perception of their own and other children's ability tends to become dichotomized when competition is stressed and there is evidence that this perceived hierarchy of ability is shared by both students and teachers. In contrast, when children focus on improving their own performance, they report significantly more effort-related thoughts and more self-monitoring. Noncompetitive cooperative classrooms tend to create a climate in which similarities rather than differences are stressed.

Ames (1987) points out that the effects of attribution retraining may be short lived when children are placed back into classrooms where competition and social comparison are the norm. The children's tendency to compare their performance with others makes it difficult for many of them to maintain a sense of self-confidence or a feeling that effort pays off. Effort becomes a double-edged sword under these conditions, because increased effort does not always lead to high levels of performance (relative to others in the class). She suggests that those who most need attribution retraining are the least likely to benefit in the long run. To maintain the positive effects of attribution retraining, classrooms must be restructured so that cooperation and a focus on self-improvement become more common.

These findings must be qualified because of *age differences* in attributions (see Nicholls & Miller, 1984). Young children do not use social comparison to the same extent. Very young children equate effort and ability and assume that those who try harder are smarter as well. They also come to school with relatively high perceptions of their own ability and rank themselves at or near the top of their class. Accuracy in social comparison tends to increase up until about age 12. This suggests that there may be a "grace" period in early elementary school when children are relatively unaffected by the negative effects of the helpless attributional style. Rholes, Blackwell, Jordon, and Walters (1980), for example, compared students in kindergarten through the fifth grade and found reduced performance and less persistence after failure only among those in the fifth grade. Only at this level were ability and effort seen as inversely related.

It is also possible that these changes reflect a gradual process of socialization in which children are taught a static view of intelligence whereby they perceive effort and ability as being inversely related. This view is widely held and leads children to conclude that really smart people do not need to try very hard and, for those who perform poorly, no amount of effort can compensate. The research on motivation and metacognition represents a frontal assault on this cultural convention and a coming together of several lines of research that have previously been unconnected. As we learn more and more about the nature of intelligence, it becomes increasingly clear that there is a dynamic interaction among culture, cognition, and motivation. The development and use of cognitive skills require effort, and attributions about the causes of success and failure help determine the extent that the person is willing to try hard in the face of potential failure. These attributions are mediated by parents, teachers, and the classroom environment, which reflect general beliefs within society as a whole.

This new awareness also makes it increasingly clear that there is no one approach to developing cognitive skills or reducing the negative effects of the attributional pattern associated with learned helplessness. One approach is to change the structure of the classroom so that cooperation rather than competition is stressed. Another is to more closely tailor instructions to each student's current level of ability. The simplest way to ensure that students come to expect success is to make sure they achieve it often by starting with relatively simple tasks and moving in small steps to more complex problems. But students must learn to cope with both success and failure, and the only way to do this is to change their beliefs about the nature of intelligence so that they can attribute failure not just to lack of effort but other potentially correctable shortcomings, such as lack of knowledge or ineffective strategies.

Children who are not metacognitively aware are *doubly handicapped* (Brown, Palincsar, & Purcell, 1986). Not only do they perform poorly, but they are unaware why their performance is poor. Teaching children specific strategies and increasing their awareness of when and where to use them shows them that differences in performance are due (at least in part) to differences in the mastery of cognitive skills. It also teaches them that the development and use of such skills requires work, but this effort pays off in increased comprehension, better memory, and more effective problem solving.

PART FOUR

Conclusion

CHAPTER 12

Implications for Social Reform

This concluding chapter examines the role of caste and class in American society. It is argued that, in addition to poverty, lower-class black Americans face a job ceiling and have a status similar to caste in other countries. Although the sources of poverty and discrimination ultimately lie in society as a whole, substantial gains may be made through early intervention and educational reform. Classroom factors that limit academic performance are examined and various alternatives to the traditional classroom are explored. The final section looks at the economic and social cost of nonintervention.

By now, it should be clear that cultural conditions influence almost every aspect of cognitive development. Society shapes intelligence both directly through motivation and cognitive skills and indirectly through various *socially sanctioned myths* about the nature of intelligence. These include the belief that intelligence is something you are born with and therefore either have or do not have, together with the corollary that, if you do not have it, it cannot be developed and education is a waste of time. A second misconception is that intelligent people do not need to try very hard and that those who do try hard are not really smart. A third belief is that intelligent people do not make mistakes and are capable of errorless performance on the first trial. And finally, there is the common belief that intelligent people are intelligent in every way, and conversely, someone who does not perform well in one area will probably do poorly in other areas as well. Each of these beliefs is widely held, and each is either greatly overstated or downright wrong. Intelligence can be improved, intelligent people do need to try hard and they do not always get it right the first time, and people can be intelligent in different ways. These beliefs have helped promote learned helplessness in those who do not perform well, and they encourage people to withhold effort or give up at the first sign of defeat. They have also made us much too willing to give up on poor students rather than take the time and effort to teach them what they need to know. A careful examination of the literature shows that no aspect of human intelligence is left untouched. Even such basic skills as speed of information processing, depth of information processing, and the capacity of working memory are strongly influenced by social conditions and previous experience. The development of metacognitive skills which are used to consciously regulate the processing and use of information teaches children that effective information processing requires time and effort and increases the probability that they will come to regard difficult tasks as a challenge rather than a source of potential failure.

Although the effects are pervasive, they do not apply to all social groups in the same way. The socialization of boys and girls, for example, is very different. Females develop attributional styles that tend to limit achievement motivation and promote the fear of failure. Males often develop a negative attitude toward school. Many lower-class males come to see school as unmasculine and turn to other activities to promote a feeling of group solidarity. These attitudes tend to limit the level of aspiration or intrinsic interest in academic tasks.

Racial differences are even more striking. Poor blacks and other disadvantaged minorities are disproportionately exposed to cultural conditions that limit the development of cognitive skills. They are far more likely to be raised in conditions that inhibit high levels of achievement motivation and foster a sense of helplessness. These motivational differences have cognitive consequences, which help determine the way people seek out, process, and use information. They may well account for the differences commonly found between blacks and whites on IQ tests. Indeed, the low performance of blacks on standard tests of "general intelligence" may be overdetermined, because it is influenced by a range of cognitive and motivational factors and by more general attitudes toward education in general.

CASTE AND CLASS IN AMERICAN SOCIETY

Although the United States prides itself on being an "open" society in which all its citizens can be whatever they want if they simply try hard enough, there are numerous forms of subtle and not so subtle discrimination and very little movement across classes. Ogbu (1986) has argued that black Americans and to a certain extent Hispanics, native Hawaiians, and American Indians have a castelike status within the United States. Blacks were brought to the United States as slaves, whereas the other groups were conquered as settlers moved west. Members of all four groups were then given menial jobs and kept in a subservient position through a combination of legal and extralegal devices. They are still frequently regarded as socially inferior by members of the white majority and are ranked as less desirable than whites as neighbors, employees, and co-workers. More important, castelike minorities often face a *job ceiling* in which it is difficult or impossible to obtain work commensurate with their skills. Blacks are far more likely to be unemployed or underemployed and earn less than whites with comparable levels of education. In 1985, the ratio of black and white unemployed high-school graduates was roughly three to one, and the proportion of unemployed black graduates was higher than that of whites who had dropped out of school (Bureau of Census, 1987).

Blacks also differ from other minorities in that they are far more likely to be isolated in hypersegregated communities (Massey & Denton, 1989). Nearly a quarter of all black Americans live in densely settled, isolated urban centers, where they rarely get a chance to interact with members of other groups. Hispanics, in contrast, are much less isolated, and there are few restrictions on native Hawaiians. Nearly 40% of blacks but only 7% of whites live in extreme poverty areas where they have little access to quality shopping areas, recreational facilities, and schools (Wilson, 1987).

Despite the tremendous legal gains and the real improvements among middle-class blacks, lower-class blacks are still stigmatized by their race and occupy something like a pariah status within the United States. Castelike minorities differ from other ethnic and religious minorities in a number of ways. Ogbu (1986) makes a distinction among autonomous, immigrant, and castelike minorities. Autonomous minorities, such as the Amish, Mormons, and Jews, are minorities in number only. They may be targets of discrimination, but they possess a distinct religious and cultural identity. Immigrant minorities, such as the Chinese, Japanese, and Vietnamese, have come to the United States to improve their social, political, or economic status. They may hold menial jobs and lack political power, but they compare themselves not with white Americans but with people back home in their countries of origin. As a result, they can see a noticeable change in their own standard of living and can expect even greater gains for their children. They also have the option of returning home or going elsewhere if they are not satisfied with the United States. Castelike minorities, in contrast, have been incorporated into the country involuntarily, relegated to low-status positions, and denied access to jobs which they are more than qualified to do.

Ogbu (1986) compares blacks in America to castes in other countries and notes some interesting parallels. These include West Indians in Britain, the Maoris or indigenous people of New Zealand, Buraku outcastes in Japan, and Harijans in India. The Buraku in Japan are particularly interesting because they and non-Buraku Japanese are members of the same race. Although they were emancipated by royal edict in 1871, this did not change the way they were perceived and treated by the dominant group. Buraku children in Japan tend to score about 15 points lower than non-Buraku children on IQ tests, and they are often one or two years behind in basic reading and mathematical skills. But when these families immigrate to the United States where all Japanese are treated more or less the same, these differences disappear.

Ogbu (1986) argues that job restrictions and lack of educational opportunities in the past have left black families ill-prepared for entry into white middle-class society. Their exclusion from high-status work has limited the development of cognitive skills associated with these professions. Although Americans in general tend to be overqualified for the jobs they possess, the degree of overqualification is particularly high among black men. High levels of unemployment are demoralizing because people look around and see a variety of people—some motivated, some lazy, some educated, some not—all without jobs. This makes it hard to believe that hard work and effort pay off. The problem is not that the black community undervalues education, but that even with education, success is difficult to achieve.

The effects of poverty and discrimination are often subtle and indirect. Infant mortality rates among blacks in 1980 was twice that of white infants. The death rate among black mothers during childbirth is also three times that of whites. This is due to a combination of early pregnancies, poverty, and lack of prenatal care. Sanders-Phillips (1989) suggests that these rates can be taken as a general indicator of the quality of health. Those who do survive may be exposed to prenatal and postnatal health hazards that do not lead to death. Racial prejudice is an "environmental stressor" (Gougis, 1986). Blacks are more likely to die from stress-related diseases and have a life span six years shorter than whites. Many stress-related problems among blacks are due to adverse social conditions, such as poverty, unemployment, and high rates of crime, but they are present even when these do not occur. High-income blacks experience more stress than high-income whites.

Poverty and low socioeconomic status also affect the way children are raised. Working-class families with jobs that lack autonomy and demand compliance tend to stress obedience in their children (Kohn, 1969) and produce lower levels of perceived control (e.g., Stephens & Delys, 1973). Middle-class mothers engage in more conversation, interact with their children more often, and provide a wider range of stimulation (Tulkin & Kagan, 1972). They also read to their children more often and this tends to promote skills that help their children in school (Brown, Palincsar, & Purcell, 1986). Middle-class and upper-class children read more books, visit museums, go to concerts, and so on and, thus, bring to school what Bourdieu (1977) has described as "cultural capital," which is highly prized by the school system. Lower-class children, on the other hand, bring a set of cultural experiences that tend to be systematically devalued by teachers and administrators.

Black children are also far more likely to be raised in single-parent homes. Nearly half of all black children live in female-headed households. Adolescents raised in mother-only homes engage in more socially deviant behavior, such as smoking, delinquency, and misbehavior in school (Dornbusch et al., 1985). These effects are particularly strong for males, and they are considerably reduced when social support is available. Giovanni and Billingsley (1970) found that parental neglect was less common in families with strong kinship ties and that a substantial amount of child abuse and mistreatment was due to a lack of social support and economic stress. Mothers who have access to social support during pregnancy reported less stress, anxiety, and depression and had a more positive attitude toward pregnancy (Tietjen & Bradley, 1982, cited in Bronfenbrenner, 1986). They also respond more quickly when their infants cry (Crockenberg, 1984) and provide more adequate child care (Epstein, 1980). Furstenberg and Crawford (1978) found that children of black teenage mothers who lived alone had more behavioral problems and scored lower on cognitive tests. Although the effects of single-parent homes are not limited to black children, they are more common in the black community because of the high frequency of teenage pregnancy, late marriages, and the tendency to not remarry after divorce or separation.

The combined effects of job discrimination, poverty, and lack of educational opportunities has led Boykin (1986) to conclude that black Americans face a *triple quandary.* They are simultaneously socialized into three somewhat conflicting roles based on mainstream American society, black culture, and their status as an oppressed minority. Despite a number of individual and regional differences, black culture has its roots in African societies, which tend to stress spiritualism, social contact, emotional expression, and a high degree of stimulation rather than materialism, egotism, and impulse control. In black society, there is an interplay between expressive individualism and ties to the community. Black culture encourages people to cultivate a distinct personality and expressive style, but is also stresses social commitment rather than individual achievement. Because black and white values are somewhat "noncommensurable," blacks often have to choose between adopting the attitudes and values of their own culture or succeeding within mainstream society. Black culture is also not recognized as a legitimate form of expression by members of the white community, who often treat it as an inferior version of their own. Cultural differences are seen as deviations from the norm and are used to justify unequal opportunities and access to work.

This leads blacks to develop various coping strategies that may limit academic achievement or the development of cognitive skills. Many black students begin to see teachers as adversaries and deliberately resist doing what they want. Others may pretend to go along but subtly undermine the teachers' efforts at the same time. Teachers begin to see these children as socially disruptive and, sometimes, as stupid. Students and teachers then actively participate in "achieving" school failure. As more and more time is devoted to discipline and control, less time is available for classroom instruction, and as a result, students in these classes begin to fall further and further behind. Blacks are three times more likely to be labeled mentally retarded or placed in special education programs (Parham & Parham, 1989). Those who do well academically are accused of acting white and often respond by shedding their cultural background and adopting the values of the white middle class. Hispanic cultures also tend to discourage academic success among females, but no other group, with the possible exception of native Americans, experiences the same degree of conflict. As DuBois (1903) expressed it near the turn of the century, "One ever feels his two-ness—an American, a Negro; two souls, two thoughts, two unreconciled strivings; two warring ideals in one dark body, whose dogged strength alone keeps it from being torn asunder" (p. 17).

EARLY INTERVENTION AND EDUCATIONAL REFORM

When one looks at the deteriorating conditions among lower-class blacks, the status of women, the plight of the homeless and the poor, it is hard not to become discouraged. It is becoming increasingly clear that cognitive development and the pattern of academic failure frequently found among poor blacks and, to a certain extent, the lower class in general have their roots in American culture and are mediated by motivational variables that limit academic success. These include the perception that jobs are limited and education does not necessarily lead to higher incomes or steady employment. Because these beliefs are based on social conditions, they cannot be changed without a program of massive economic reform, which should include job creation, reconstruction of the inner cities, family planning, drug rehabilitation, and prison reform.

It is not altogether clear that Americans are willing to pay for the cost of economic reform. Many continue to blame blacks and hold them personally responsible for their low social and academic status. Poor students are seen as disruptive or stupid, and they, therefore, deserve to fail. Some believe that they can continue to enjoy their wealth by avoiding black areas and buying new locks or security systems, whereas others cling tenaciously to the belief that government intervention is not necessary because wealth will trickle down. Many Americans have lost a clear vision of what America should look like in 20 or 30 years, and this lack of vision creates impotence in the face of pressing social problems.

Given these limitations, a less controversial starting point is *educational reform*. Nobody believes that educational reform will cure the United States of social injustice, but schools seem to be a good place to begin. Many Americans agree that the current educational system is not meeting the needs of students and is no longer adequate. In 1983, the National Commission on Excellence published a study called *A Nation at Risk,* in which it warned that the educational foundation of American society is being eroded by "a rising tide of mediocrity." It went on to say "If an unfriendly foreign power had attempted to impose on America the mediocre educational performance that exists today, we might have viewed it as an act of war. As it stands, we have allowed this to happen to ourselves" (p. 5).

Schools are also one of the few major institutions directly in the public domain, and they can, therefore, profit more directly from government programs. Educational opportunities vary widely, but it may be possible to reallocate resources so that they go where they are most needed. The current system, in which local communities finance schools, has provided good suburban schools and substandard schools within the inner cities. As a result, poor children enter school without a level playing field, and schools have served as a primary means by which existing status differences are passed on from generation to generation (see Bowles, 1977).

Much of the previous emphasis has been devoted to offsetting the presumed negative effects of the early home environment by providing *preschool programs.* It was assumed that poor school performance among lower-class children was due to conditions occurring prior to school, such as a lack of stimulation or less interaction between parents and children. As a result, preschool programs have typically taken one of three forms (see Darlington, 1986). One approach is to provide nursery school programs for 3- and 4-year-olds, in which children receive instruction either on a one-to-one basis or in small groups. A second approach consists of home-based programs, in which poor families receive toys, activities, and games; and mothers learn how to use them and how to interact with their children so as to promote academic success. The third approach combines nursery school programs with periodic home visits.

Early assessments of these programs, such as the 1969 Westinghouse Report on Project Head Start, were quite negative. These studies found that there were some initial gains in vocabulary and IQ but these soon faded. By the end of the second grade, the test scores of those in the Head Start Program were not significantly different from those of similar children without preschool training. This led Jensen (1969) to conclude that "Compulsory education has been tried and it apparently has failed" (p. 69).

More recent longitudinal research, on the other hand, suggests that Head Start may have had a *delayed effect* (e.g., Consortium for Longitudinal Studies, 1983). Darlington (1986) has provided a review and has found that, by high school, those who received preschool training were doing much better than comparable peers. They were half as likely to have been placed in special education classes, much less likely to repeat a grade, and more likely to finish school. Neisser (1986b) suggests that these differences are not due to any specific content but to a change in attitude—"For one year at least, the schools took the children (and their parents) seriously: responded thoughtfully to their behavior, believed in their potential achievement, listened to what they had to say" (p. 11). The increased attention gave students and their parents an extra measure of confidence and provided them with the persistence necessary to continue with their schoolwork when conditions became difficult.

As we gain more and more knowledge about the role of motivation and the nature of cognitive development, it becomes possible to create conditions that enhance education and provide procedures specifically designed to overcome existing deficiencies. These innovations are not limited to preschool. They can be applied at any time. They work well because they are based on a clearer understanding of underlying processes and are often closely tailored to each student's current level of ability.

Some of the research on educational reform is hampered because many attempts to improve the quality of education are problem centered rather than research oriented. Many people have been willing to try anything and everything as long as it works, but even the unsystematic research helps show that significant gains can be made. James Comer (1989), for example, describes a program carried out by the Child Study Team at Yale University that involved two low-income, black elementary schools in New Haven. When the study began in 1969, these schools ranked last and next to last in achievement, were 1½ years behind in language and mathematical skills, and had low attendance. One of the original schools dropped out of the project and was replaced by a similar school. Fifteen years later, with no change in socioeconomic conditions, the two remaining schools ranked third and fourth among the 26 elementary schools in the area, were performing about a year ahead of grade, and ranked one and two in attendance. These changes were based on a number of factors, including more effective planning, staff development, and greater parent-teacher cooperation, but they demonstrate that dramatic gains can be made without changing structural conditions or the economic position of blacks within the local community.

A more systematic study was carried out by Ronald Edmonds (1979, 1986), who compared effective and ineffective schools. Effective schools were defined as those who had a similar proportion of rich and poor students who exceeded a minimum standard of academic achievement. Such schools were rare, but Edmonds found that two similar-size schools in the same district with similar financial resources, racial and socioeconomic compositions, and student-teacher ratios could differ substantially. One of the key variables was *leadership*. Effective schools had principals who were strong leaders and who had clear goals about where the school should be going. Irvine (1990) points out that effective principals are hardworking and dedicated and have interpersonal skills that facilitate face-to-face contact with a variety of people. They tend to be extroverted and listen unpatronizingly to parents, teachers, and students. Although they set clear goals that teachers and other administrators generally understand, they are willing to ignore bureaucratic rules and regulations when they are incompatible with the school's needs. They are also good teachers who can observe classroom interactions and provide teachers with specific feedback. Because they know how to teach, effective principals are more skilled at recruiting good teachers and reassigning those who are incompetent.

Teachers affect school performance in many ways. Several metaanalyses of teachers' expectancies suggest that teachers prefer students who are white, physically attractive, and middle class (Baron, Tom, & Cooper, 1985; Dusek, 1985). Blacks are criticized more than whites, and boys receive more criticism than girls (Irvine, 1985, 1986). Teachers also use harsher language when criticizing boys. The effects of these differences are often subtle and indirect. When teachers become angry, children often attribute their poor performance to insufficient effort, but when teachers show sympathy, lack of ability is assumed (Graham, 1988). Because girls receive less criticism and because it is based primarily on academic performance, they tend to take it more seriously and see it as a sign of low ability (Dweck, 1975). Many students who are not liked by their teacher come to dislike themselves. This is especially true for low-income students because they tend to hold their teachers in high regard (see Irvine, 1990).

Other problems occur because students and teachers come from different cultural backgrounds. There has been a significant decline in the number of teachers from minority groups. In 1950, half of all black professionals were teachers. By 1980, minority teachers constituted 12% of the total work force, but the proportion has dropped to about 5% in recent years (Smith, 1988). This means that the average student will have only about two minority teachers prior to college. Many black teachers now come from middle-class backgrounds or from areas that are isolated from lower-class black communities where remnants of black culture are most strong.

As a result, middle-class teachers and lower-class black students frequently experience what Jacqueline Irvine (1990) has called a "lack of cultural synchronization" that leads to misunderstandings and inappropriate responses. Black culture, for example, encourages a preference for relatively high levels of stimulation. As a result, black boys from lower-class communities are often more boisterous, and verbal sparring frequently turns into rough-and-tumble play, which intimidates white teachers and students. When black children express themselves with great energy, teachers often see them as overly aggressive. Lower-class black children come to school less prepared to answer the kind of open-answer questions that prevail in the classroom. Because black children are more critical of experts and published works, teachers often see them as illogical and unwilling to listen to reason. Blacks hesitate to share personal information and expect authority figures to act with authority. They are, therefore, bewildered and confused when teachers try to act like friends. Their reluctance to maintain eye contact causes many teachers to assume that they are not paying attention. On entering school, lower-class black children often discover that the ground rules have completely changed, and this lack of synchronization increases rather than decreases when teachers pretend not to notice racial differences and treat all children the same.

Similar problems occur during *parent–teacher interactions* (Irvine, 1990). Black parents overwhelmingly support high academic standards, tough courses, and more homework, because they understand that education is one of the few ways for their children to break the vicious cycle of poverty (Honig, 1987). Conflicts between black parents and teachers are not due to disagreement about the value of education but a lack of mutual understanding. Schools often treat black parents as a nuisance and hold them personally responsible for their children's disruptive behavior or lack of academic success. Many black parents understand this attitude and avoid confrontation because they believe their children will be punished by teachers who are questioned or criticized. This lack of participation is taken as proof that poor black parents are not interested in their children's education.

Factors that limit parental participation increase as children grow older. Adolescents often discourage their parents from participating because they believe their friends will disapprove. Slaughter (1986), for example, found that black students whose parents were involved were less liked by their peers. Parents of high school students are less likely to offer help or question their children about course work because the material is more difficult. A further barrier is the sheer number of high school teachers who must be contacted.

What is so striking about these misunderstandings is that virtually nothing is as it seems. Lack of eye contact or failure to respond is taken as a sign of disinterest or stupidity. Lack of parental participation is seen as proof that parents do not care. It is as if these interactions were being orchestrated by an "unseen hand," and to an extent, they are. Members of each culture attempt to understand the other from their own perspective and are often totally oblivious to the underlying dynamics that are the source of these misperceptions.

This lack of cultural sync leads to hidden conflicts, hostility, and a breakdown in communication, but it can be offset to some extent by a better understanding of black culture. The declining number of black teachers means that lower-class black students will be increasingly exposed to white teachers who may not understand them. Two options are possible. Courses in black culture (and those of other minorities) could be made a mandatory part of teacher education, or teachers could be given additional incentives for taking courses in black history and culture. Similar courses should be made a fundamental part of the high school curriculum, and they should include an open discussion of racial discrimination in the United States. By providing students with an accurate assessment of the opportunities available, it may be possible to overcome the negative effects associated with self-blame and feelings of inferiority. An accurate description of socioeconomic conditions could also serve as a useful means for measuring progress (if it does occur).

The role of parents and teachers in promoting academic achievement is particularly strong for lower-class blacks. Clark (1983) studied 10 poor black families in Chicago with high- and low-achieving children and found that families with academically successful children encouraged independence, talked with their children more often, set clear guidelines, and had high expectations that their children would succeed. They also visited their children's school more often and helped with their homework. Those with low-achieving children had low expectations, seldom visited their children's school, offered little support, and did not expect their children to attend college. Effective teachers also have high expectations and do not prejudge students on the basis of race, family income, or test scores. Manns (1981) had black and white adults from lower-class and middle-class backgrounds who had succeeded academically describe the perceived cause of their success and found that low-income blacks reported more instances of social support from some significant other. Relatives were mentioned most often, but the majority of unrelated supporters were teachers.

One practice that has been particularly detrimental to poor and black students is *tracking*. When students are assigned to groups on the basis of perceived ability, the number of low-income and black students in low-track groups is disproportionately high. Tracking is quite common and often occurs in the early school years. Rist (1970), for example, found that children in kindergarten were separated into groups on the basis of cleanliness and dress. These divisions are later used by teachers in elementary school to assign children to reading groups. Although the goal of tracking is to tailor education to individual students, teachers often determine the number and size of groups before actually meeting the students and there is very little mobility once the decision is made.

Once children have been assigned, their instruction varies, depending on the track (Brown, Palincsar, & Purcell, 1986; Collins, 1986). Students in low reading groups concentrate primarily on decoding and pronunciation and come to see the purpose of reading as pronouncing words correctly without making mistakes. Analysis of their reading suggests that they read words item by item, almost as if they are reading from a list. Good readers are questioned more about the meaning of a passage and are frequently asked to criticize or evaluate the material. They spend more time actually reading, and errors are tolerated as long as the material is understood. Because the fundamental skill of reading determines performance in other areas, those assigned to low reading groups build up handicaps that become increasingly difficult to overcome.

Whereas groups are ranked initially to increase the quality of instruction, a kind of institutional inertia sets in once these groups have been formed. There is seldom movement from one group to another, and when it does occur, children are far more likely to move down than up (Rosenbaum, 1976). Estimates of change vary from 9% (Hawkins, 1966) to 16% (Groff, 1962). Some schools even track by assigning children to different classrooms and when they do, changes are even more rare. Jackson (1964) studied across-classroom groupings and found that only 1% to 5% of the children changed classes within a year. This means that early decisions, sometimes made on the basis of performance *in kindergarten,* may have lasting effects and that many of those assigned to lower groups are virtually written off at age five.

Tracking continues throughout high school, where it is used to assign students to academic and nonacademic programs. Income and race seem to play a key role. Jones, Van Fossen, and Spade (1987), for example, found that two-thirds of high ability, middle-class students were in academic programs, but only half of lower-class students with high ability were in the academic track. Although achievement tests are often used to assign students to programs, Gilmore (1985) suggests that teachers' perception of black students' attitude may be more important than academic ability. Teachers and administrators are not always conscious of the basis for their decisions, but as Irvine (1990) points out, assigning black students to low tracks—either intentionally or inadvertently—produces a phenomenon known as resegregation. Resegregation separates black and white students and results in two separate schools—one black and one white—together but unequal in the same building. The high frequency of black students in low-ability groups reinforces the stereotype that blacks are less gifted than whites.

The practice of tracking tends to reinforce existing status differences, legitimize unequal treatment, and maintain the status quo. Lower track groups are often avoided by more experienced teachers, and the methods of instruction tend to be less effective and may contribute directly to disruptive behavior. Teachers and guidance counselors sometimes use subtle techniques to channel students into particular tracks but do so in such a way that both parents and students believe the choice was their own (Rosenbaum, 1976). The existence of a vocational curriculum helps solidify a vague, ill-defined preference for manual labor among lower-class children by pressuring them to select nonacademic programs before they enter high school so that they can be with their friends (MacLeod, 1987). Black students are more likely to be enrolled in general or vocational programs and take fewer academically rigorous courses. Only 33% of black students, as opposed to 40% of white and 52% of Asian students, are in college preparatory classes (The College Board, 1985). Those taking vocational courses often find their diplomas have little value because the skills taught and equipment used in high school are often outdated or even obsolete.

The competitiveness of the traditional classroom widens the gap between good and poor students. Ames (1987) and her associates have consistently found that children who compete with each other tend to compare their performance and make ability attributions for success and failure. Those who perform well come to believe that they have high ability, whereas those who are less successful come to assume that they are not as smart. Students' judgments of their ability and that of their classmates tend to polarize when competition is stressed, and there is substantial agreement about who is smart and who is not. In competitive classrooms, success is relative and one child's good performance makes others look bad. Only a handful of students thrive under these conditions, and those that do not may find it difficult to concentrate, monitor their performance, or maintain an interest in school. Because people tend to devalue those things they do poorly, poor students gradually lose interest in school, become restless, and channel their energy into other activities.

Noncompetitive, cooperative classrooms, in contrast, create a climate in which similarities rather than differences are stressed. When children focus on improving their performance rather than competing with others, ability attributions are rare. Children in noncompetitive situations report significantly more effort-related attributions and engage in more goal-directed thoughts. Children in noncompetitive classrooms compare their current level of performance with their previous performance, and since most do improve, it is relatively easy for them to maintain the belief that they are doing well. Perhaps more important is that children attribute these improvements to the degree of effort or the development of new skills. When students focus on improving performance, they tolerate errors and mistakes more easily and become more willing to take risks and confront challenges.

The growing recognition that traditional approaches to education are severely limited has led to the development of a variety of alternative approaches, some of which have already been mentioned. These include cooperative learning, reciprocal teaching, peer tutoring and direct intervention for those with specific problems. *Cooperative learning methods* have been developed by Slavin (1982, 1987). This approach divides the classroom into four- or five-member teams composed of students who currently perform both poorly and well. Teams typically consist of both boys and girls and, where possible, students from different racial and ethnic groups. Team members work on projects together until all members of the group are ready to be tested. Students are then tested individually and a team score is calculated. A variation of this approach uses a similar procedure, but teams compete in weekly tournaments. Cooperative learning not only increases the performance of black students, it has been found to improve race relations as well (Slaven & Madden, 1979).

The approach known as *reciprocal teaching* has already been described. Palincsar and Brown (1984) have used the procedure to significantly increase the reading ability of poor readers by teaching them to ask questions, clarify difficulties, predict what will happen, and summarize the material. It can be used either with individual students or small groups. The teacher begins by modeling and describing the desired behavior, being careful to closely tailor the discussion to the students' current level of ability. Students are told why the procedures are useful and when they should be used. Students then repeat the activity and the interchange goes back and forth until the material is mastered. Reciprocal teaching makes underlying processes overt, explicit, and concrete. By customizing the material to each student's current level of ability, students can proceed step by step until the skills are acquired. Students learn that mistakes are a natural part of the learning process but that they are often due to a lack of skill and are therefore not insurmountable. Similar procedures have been used to teach children to listen more carefully, write, tell stories, and solve problems (see Gelman & Brown, 1986).

A big advantage of reciprocal teaching is that those who have been trained gain the explicit knowledge necessary to teach others. They can then serve as *peer tutors* for those who are not doing well. Peer tutors can be either trained or selected for their high level of ability, but the procedure serves a dual function. Poor students gain knowledge and skills from the close one-on-one contact, but the tutors also increase their own explicit knowledge and get a chance to practice the procedures until they become more and more automatic. Unlike competitive classrooms where one student's success has negative implications for others, peer tutoring creates a situation in which everybody wins. By combining groups with different ethnic and racial backgrounds, peer tutoring can also break down barriers that keep people and groups apart.

As more and more is learned about cognitive development and the underlying processes that distinguish those who perform well in a given area, it becomes possible to provide very specific instructions to those who do not perform well. A number of examples have already been described. People have been taught to solve problems more effectively by recognizing the biases and distortions that typically occur, special procedures have been developed to teach mathematics and scientific concepts through computer programming, and there is a growing body of knowledge on improving performance by teaching metacognitive skills. Glaser (1984, 1989) points out that there is a bit of a trade-off in the two most common forms of instruction. One approach stems from the research comparing experts and nonexperts and focuses on domain-specific knowledge. The other is derived from the research on metacognition and attempts to teach higher order executive processes that can be used in a variety of domains. Domain-specific knowledge is more easily taught, but it may not transfer to other domains. He suggests that a third alternative is to teach metacognitive and domain-specific skills at the same time by describing metacognitive processes within the context of a particular domain.

Although the present chapter has focused primarily on differences between black and white students, other groups face formidable obstacles that limit academic achievement. One of the most common is the language difficulties experienced by those whose native language is not English. The number of language-minority students has increased substantially in recent years. Current estimates place the number somewhere between 1.5 and 2.6 million. About 75% of these are Hispanic. There is a movement to limit the use of other languages in school, but current research suggests that these attempts are misguided (see Padilla et al., 1991). Crawford (1989), for example, compared English immersion programs with transitional programs designed to introduce English quickly or gradually over a period of several years. Students who were introduced to English gradually scored higher on almost every academic subject, including English. Those submitted to sink-or-swim immersion programs, in contrast, had lower levels of academic achievement and were less proficient in English. Research has also shown that people who are bilingual enjoy other cognitive advantages, such as higher levels of cognitive flexibility, metalinguistic awareness, concept formation, and creativity (Hakuta & Garcia, 1989). Difficulties with English can sometimes produce surprising results. Asian Americans, for example, experience a great deal of difficulty with English and score lower (as a group) than whites on the verbal section of the Scholastic Aptitude Tests. As a result, they tend to concentrate on subjects requiring quantitative skills, such as mathematics, computer programming, and science, where they do exceptionally well (Sue & Okazaki, 1990).

A final problem with the current educational system is an overreliance on standardized tests to select people for particular programs or stream them into particular tracks. The presence of timed tests is an administrative convenience that tends to penalize those who do not speak English well and those who respond slowly. The strong correlation between metacognition and effective problem solving challenges the traditional notion that intelligence is based on speed. Sternberg (1985a) has pointed out a number of areas in which speed and effectiveness are inversely related. The first is based on differences between reflective and impulsive cognitive styles. Reflective people carefully consider various alternatives before they make a decision, whereas impulsive people respond quickly and are more likely to be wrong. Sternberg's (1977, 1981) own work on problem solving shows that "more intelligent" people spend more time planning strategies and encoding individual items. A study of insight learning carried out by Sternberg and Davidson (1982) in which people were free to spend as much time as they wanted found a high positive correlation between time spent and measured IQ. Finally, a study by Wagner and Sternberg (cited in Sternberg, 1985a) found that more skilled readers spend more time reading passages when they believe they will be tested in detail. Research comparing people with an internal and external locus of control has also found that internals generally spend more time seeking information and making decisions (see Chapter 4).

Sternberg (1985a) points out that it would be foolish to assume that speed is never important, but most of the problems occurring in day-to-day life require careful consideration and persistence rather than a quick response. Those who support the notion that speed is important claim that the ability to perform individual operations quickly determines intelligence, but even when components are broken down, effective performance is associated with speed in some cases and slowness in others. Verbal comprehension, effective problem solving, and creativity all depend on higher order executive processes that take time to be used effectively. The widespread use of so-called aptitude tests and achievement tests to place people in tracks and screen candidates for law, medical, and graduate programs may systematically eliminate many of those who are most qualified.

Even a brief examination of educational research shows that there is no single solution to the current educational crisis. A variety of alternatives need to be considered. As Irvine (1990) has pointed out, the "typical" student of the past, who was highly motivated, achievement oriented, white, and middle-class is becoming increasingly rare. Declining fertility among white middle-class women and increasing fertility among minority groups, as well as the influx of immigrants from Mexico and Asia, will completely alter the demographics of American schoolchildren by the turn of the century. These factors, if ignored, will contribute to a growing number of school failures, which will reverberate throughout society as a whole.

THE SOCIAL COST OF NONINTERVENTION

The current concern with educational reform stems partly from the belief that equal access to a quality education can help offset to a certain extent current inequalities based on poverty and unemployment. The real irony is that blacks as a group have made substantial gains in education, but this has had little impact on their overall socioeconomic status. The Bureau of Census (1987, 1990) figures show that more blacks are staying in school, graduating, and going on to university. The dropout rate among blacks declined from over 22% in 1970 to about 12% in 1988, and it is now virtually the same as that of whites. The proportion of blacks in university is almost as high as that of whites and, in the mid-1980s, it was actually higher. Blacks as a group also have significantly higher levels of education than other minorities, such as Hispanics or American Indians.

Despite these impressive gains, blacks still earn about 67% of what whites make and about 20% less than Hispanics. This discrepancy between blacks and other groups occurs at every level of education. The average white worker with a college education earned over $50,000 a year in 1987, but the average college-educated black earned less than $38,000. The median income of black families is still less than 60% of the amount earned by white families, and this figure has remained virtually unchanged for the past 20 years. The rate of unemployment among blacks has actually increased during the past 15 years, and it has hit educated blacks particularly hard. In 1985, black high school dropouts were twice as likely to be unemployed as white high school dropouts (51% versus 28%) but, among graduates, the ratio was three to one (34% versus 12%). Education still increases the chance of finding work but the rate of unemployment among black high school graduates is higher than that of whites who have dropped out of school, and the average income is considerably less.

Other indirect indicators also show a lack of progress. Black children are actually losing ground. The number of black children born out of wedlock is increasing and they now account for more than 60% of all black births. There has been a steady increase in the number of single-parent homes, and nearly half of all black children currently live below the poverty line. Despite gains in some areas, such as the number of black elected officials and the number of black-owned businesses, blacks continue to have high rates of infant mortality, maternal deaths, juvenile delinquency, and crime. The average life expectancy has been increasing, but it is still about six years less than that of whites. The life expectancy of a black man in Harlem is less than that of an adult male in Bangladesh.

The stubborn persistence of social problems and the failure to make economic gains in spite of increases in education suggest that educational reform may not be sufficient. It is not that education does not matter. Highly educated blacks are less likely to be unemployed, and when they do find work, they receive higher salaries than blacks with less education. This is true for all groups. The rate of employment and the average income both increase with the level of education. But blacks as a group lag far behind whites and Hispanics with a similar level of education.

Braddock and McPartland (1987) have recently surveyed a national sample of over 4,000 employers and have uncovered several reasons why blacks profit less from education. They found that the two most common forms of recruitment for low-status jobs were informal referrals and unsolicited walk-ins. When a job becomes available, current employees are told about the position and they pass the information along by word of mouth. Blacks frequently lack access to these informal networks, and they are, therefore, less likely to know when jobs are being filled. Educational credentials play a very small role. What is more important is the ability to make a good impression and get along. Although upper level positions are more likely to be advertised, informal networks play a major role in recruitment as well.

Once people apply for work, a major criterion used to screen candidates is previous work experience. Because unemployment is particularly high among black youth, they are less likely to have prior employers who can submit letters of recommendation. This creates a vicious cycle in which unemployment early in life leads to high rates of unemployment later on. When blacks do find work, they are more likely to have low-paying jobs. Informal networks and referrals are used because they save time and money. Although they inadvertently discriminate against blacks, employers are not likely to abandon their current practices without some kind of pressure or incentive.

These problems persist because they fall disproportionately on members of the black community. They can, therefore, erroneously be attributed to blacks themselves rather than to social and economic conditions. Many whites continue to believe that blacks are personally responsible for their low status because they are undermotivated, less intelligent, or inherently lazy. These beliefs tend to legitimize current status difference, prevent constructive social change, and maintain the status quo. Unfortunately, some blacks come to share these beliefs and lose confidence. They become disenchanted with the educational system and are, therefore, willing to accept low-paying, low-status jobs.

The lack of economic progress among blacks suggests that more radical measures may be necessary. One possibility might be a job creation program similar to the Public Works Administration during the Great Depression. This should include a concerted effort to rebuild the inner cities and provide low-cost housing for people in need. One of the most visible changes in American society is the emergence of a new group of people who are *homeless*. Nobody knows precisely how many people are homeless, but current estimates range from a quarter of a million to as many as three million. The General Accounting Office suggests that the figure is increasing annually at a rate somewhere between 10% and 38%. Rossi (1990) has pointed out that there are significant differences between the old and new homeless. The "old homeless" of the 1950s were predominately older, alcoholic, white men. Most had no families, but they were not literally homeless because 80% rented rooms in cheap hotels (Bogue, 1963). Many were employed either full or part time, and others drew Social Security benefits. The old homeless were also relatively invisible since they were concentrated in rundown areas and were not allowed to sleep in doorways or public places in more prosperous parts of town.

The new homeless, in contrast, are much younger, poorer, and more destitute. There has been a significant increase in homeless women and families and a far greater proportion of the new homeless are blacks, Hispanics, or members of other minorities. The new homeless are more visible—they can be found sleeping in doorways, cardboard boxes, and abandoned cars. Very few have either full or part-time jobs. The similarities between the old homeless and the new homeless, however, are a high rate of alcoholism, mental illness, and physical disabilities. Estimates of the rate of mental illness among the homeless vary widely, but most studies suggest that the figure is close to 30%. Poor health also plays a major role and approximately 40% have some kind of physical disorder. The average life span of the homeless is only slightly more than 50 years.

In addition to housing and unemployment, a third problem confronting the black community and the United States as a whole is the increasing number of black children born into single-parent homes. The number of black children born out of wedlock and raised in single-parent homes is increasing rather than decreasing, and these trends are important because children raised by single mothers are more likely to grow up in poverty and be persistently poor. This condition is made worse because the old pattern of social support based on the extended family is also becoming increasingly rare (Wilson, 1987). Mothers with illegitimate children are also more likely to have daughters who become unwed mothers, thus creating a vicious cycle of poverty that is passed from generation to generation.

Financing job creation programs, public housing, and family planning would cost money, but these costs would be offset, to a certain extent, by a decrease in welfare payments, unemployment, and crime. Brenner (1984), for example, carried out a study for the Joint Economic Committee of the U. S. Congress that suggested a 10% increase in unemployment in 1980 would have:

1. Increased the total number of deaths in the United States by 1.2% (24,250 deaths).
2. Increased suicides by 0.7% (189 deaths).
3. Increased mental hospital admissions by 4.2% (5,885 new patients).
4. Increased arrests by 4% (403,830 additional arrests).
5. Increased fraud and embezzlement by 3.4% (8,078 additional cases).
6. Increased assaults by 0.8% (4,919 additional assaults).

Brenner noted that the effects of unemployment fall disproportionately on poor young people and on members of minority groups, but they spill over into other groups. They represent the *hidden cost of unemployment* and they give some indication of what to expect if unemployment within the black community remains extremely high.

Many people are reluctant to address these issues and hope that the problems can be contained. The current high rate of social deterioration and crime are largely limited to lower-class communities, but they are already beginning to affect people in other parts of town. In some cities, criminals from poor areas are beginning to commute into wealthy areas where high-priced commodities are more readily available. Border clashes between neighboring communities are quite common, and wealthy people are increasingly becoming victims of brutal and seemingly senseless crimes. As frustration builds up, crimes of violence become more common. Stopgap measures, such as safe communities where upper-middle-class families are protected by armed guards and high walls, are not going to solve the problem in the long run. Fortress America is not likely to hold.

In a time of soaring deficits and financial restraints, it is unlikely that affluent Americans will be willing to dig deep in their pockets and pay for either the cost of high-quality education or economic reform. Many conservative social critics, and people in general, are understandably reluctant to spend additional money given the track records of previous administrations. Previous attempts to improve economic and educational opportunities have often failed because those initiating the changes have failed to recognize that social conditions are based on an *interaction* between individuals and society. These failures have often taken one of two forms.

Those who start from the "psychological" perspective often assume that social problems and conflicts are due to mistaken attitudes and beliefs. Racial groups do not understand each other because they are uninformed (or misinformed). People give up too easily because they have come to believe that they have little control over their own lives. The key to social reform in each case is to change these beliefs so that people gain a sense of mutual understanding and develop a sense of personal responsibility. What this approach fails to consider is that these beliefs are themselves a product of social conditions and are not likely to be substantially altered as long as the social conditions that promote them are maintained.

Others have assumed that attitudes and beliefs can be more or less ignored and that constructive social change occurs through improving concrete social conditions. They argue that the lives of poor people can be substantially improved by providing better food, housing, and education. This approach has often failed because it does not take people's *perception* of the situation into account. Unemployed people have not responded positively to public housing, for example, because it tends to isolate them and break up old communities. Educational reform is also not likely to occur by improving the curriculum, if black children see teachers as the enemy and the educational process as a conflict between those trying to superimpose their own dominant white values and those trying to resist. Under these conditions, passive and not so passive resistance may seem to be the most effective strategy, and so-called troublemakers may be seen as role models to be admired.

The failure to understand the perspective of those being helped also leads to the paradox that social workers and others engaged in social reform are often seen as paternalistic do-gooders who do not understand what is going on and are despised by the very people who need them the most. Young people who enter these professions with the best of intentions come to sense this hostility and become disillusioned and cynical. They develop an us-versus-them attitude that mirrors the dominant attitude in society as a whole, but it gains credibility because it is based on personal experience.

Although the failure of both approaches seems to imply that social events are chaotic, unpredictable, and not subject to social control, such attempts have foundered because of their one-sidedness. What is needed is a concerted effort to improve social conditions and a deeper understanding of how these changes are perceived by the local community. Such an approach would have to take a larger perspective and incorporate the insights of many previously separate disciplines.

Educational reform is probably the least controversial form of social change, but education by itself is unlikely to solve the United States' most pressing social problems or eliminate the current status differences between blacks and whites. What is really needed is a more substantial package of economic and social reform coupled with a better understanding of how these changes are perceived. The cost of intervention is high, but the human cost of nonintervention may well be staggering. Current conditions within the inner city may be just a taste of things to come. If these problems are not addressed, they may get worse until, for many, the American dream becomes a nightmare.

References

Abramson, L. Y., Seligman, M. E. P., & Teasdale, J. D. (1978). Learned helplessness in people: Critique and reformation. *Journal of Abnormal Psychology, 87,* 49–74.

Adorno, T. W., Frenkel-Brunswik, E., Levinson, D. J., & Sanford, R. N. (1950). *The authoritarian personality.* New York: Harper.

Allport, G. W. (1937). *Personality: A psychological interpretation.* New York: Holt, Rinehart and Winston.

Alper, T. G. (1974). Achievement motivation in college women: A now-you-see-it-now-you-don't phenomenon. *American Psychologist, 29,* 194–203.

Alter, J., Brailsford, K., & Springen, K. (1988, March). Why we can't wait any longer. *Newsweek,* 42–43.

Ames, C. (1987). The enhancement of student motivation. *Advances in Motivation and Achievement, 5,* 123–148.

Anastasi, A. (1958). Heredity, environment, and the question "How?" *Psychological Review, 65,* 197–208.

Anderson, A. M. (1982). The great Japanese IQ increase. *Nature, 297,* 180–181.

Anderson, C., & McGuire, T. (1978). The effect of TV viewing on the educational performance of elementary school children. *The Alberta Journal of Educational Research, 24,* 156–163.

Anderson, D. R., Alwitt, L. F., Lorch, E. P., & Levin, S. R. (1979). Watching children watch television. In G. Hale & M. Lewis (Eds.), *Attention and the development of cognitive skills.* New York: Plenum.

Anderson, D. R., & Collins, P. A. (1988). *The impact on children's education: Television's influence on cognitive development.* Washington, DC: U.S. Department of Education, Office of Educational Research and Improvement.

Anderson, D. R., Lorch, E., Field, D., Collins, P., & Nathan, J. (1986). Television viewing at home: Age trends in visual attention and time with television. *Child Development, 57,* 1024–1033.

Andrisani, P. J., & Nestel, G. (1976). Internal-external control as contributor and outcome of work experience. *Journal of Applied Psychology, 61,* 156–165.

Angyal, A. (1941). *Foundations for a science of personality.* New York: Commonwealth Fund.

Arkin, R. M., & Baumgarden, A. H. (1985). Self-handicapping. In J. H. Harvey & G. Weary (Eds.), *Attribution: Basic issues and applications.* New York: Academic.

Arnheim, R. (1962). *Picasso's Guernica: The genesis of a painting.* Berkeley, CA: University of California Press.

Asarnow, J. R., & Meichenbaum, D. (1979). Verbal rehearsal and serial recall: The mediational training of kindergarten children. *Child Development, 50,* 1173–1177.

Asch, S. E., & Witkin, H. A. (1948a). Studies in space orientation: I. Perception of the upright with displaced visual fields. *Journal of Experimental Psychology, 38,* 325–337.

Asch, S. E., & Witkin, H. A. (1948b). Studies in space orientation: II. Perception of the upright with displaced visual fields and with body titled. *Journal of Experimental Psychology, 38,* 455–477.

Astin, H. (1974). Sex differences in mathematical and scientific precosity. In J. Stanley, D. Keating, & L. Fox (Eds.), *Mathematical talent: Discovery, description, and development.* Baltimore, MD: Johns Hopkins University Press.

Atkinson, J. W. (1957). Motivational determinants of risk-taking behavior. *Psychological Review, 64,* 359–372.

Atkinson, J. W. (Ed.). (1958). *Motives in fantasy, action, and society.* Princeton, NJ: Van Nostrand.

Atkinson, J. W. (1974). Motivational determinants of intellective performance and cumulative achievement. In J. W. Atkinson & J. O. Raynor (Eds.), *Motivation and achievement.* New York: Wiley.

Atkinson, J. W. (1983). *Personality, motivation, and action.* New York: Praeger.

Atkinson, J. W., & Feather, N. T. (Eds.). (1966). *A theory of achievement motivation.* New York: Wiley.

Atkinson, J. W., & Litwin, G. H. (1960). Achievement motive and test anxiety conceived as motive to approach success and motive to avoid failure. *Journal of Abnormal and Social Psychology, 60,* 52–63.

Atkinson, J. W., & Raynor, J. O. (Eds.). (1974). *Motivation and achievement.* Washington, DC: Winston.

Atkinson, R. C., & Shiffrin, R. M. (1968). Human memory: A proposed system and its control processes. In K. W. Spence & J. T. Spence (Eds.), *The psychology of learning and motivation* (Vol. 2). New York: Academic.

Austrian, R. W. (1976). *Differential adaptation of field independent and field dependent subjects to therapy-analogue situations varying in the degree of structure.* Unpublished doctoral dissertation, New York University, New York.

Backman, J. G. (1970). The impact of family background and intelligence on tenth grade boys. *Youth in transition.* Ann Arbor, MI: Institute for Social Research.

Baddeley, A. (1982). *Your memory: A user's guide.* New York: Macmillan.

Bagrow, L. (1948). Eskimo maps. *Imago Mundi, 5,* 92–93.

Bahrick, H. P. (1984). Semantic memory content in permastore: 50 years of memory for Spanish learned in school. *Journal of Experimental Psychology, 113,* 1–29.

Bahrick, H. P., Bahrick, P. O., & Wittlinger, R. P. (1975). Fifty years of memory for names and faces: A cross-sectional approach. *Journal of Experimental psychology: General, 104,* 54–75.

Bandura, A. (1977). Self-efficacy: Toward a unifying theory of behavior. *Psychological Review, 84,* 191–215.

Bandura, A. (1986). *Social foundations of thought and action: A social cognitive approach.* Englewood Cliffs, NJ: Prentice-Hall.

Bandura, A., & Schunk, D. H. (1981). Cultivating competence, self-efficacy, and intrinsic interest through proximal self-motivation. *Journal of Personality and Social Psychology, 41,* 586–598.

Banks, J. A. (1984). Black youth in predominantly white suburbs: An exploratory study of their attitudes and self-concepts. *Journal of Negro Education, 53,* 3–17.

Barker, S. F. (1965). *The elements of logic.* New York: McGraw-Hill.

Baron, J. (1981). Reflective thinking as a goal of education. *Intelligence, 5,* 291–309.

Baron, J., Badgio, P. C., & Gaskins, I. W. (1986). Cognitive style and its improvement: A normative approach. In R. J. Sternberg (Ed.), *Advances in the psychology of human intelligence.* Hillsdale, NJ: Erlbaum.

Baron, R. M., Tom, D. Y., & Cooper, H. M. (1985). Social class, race and teacher expectations. In J. B. Dusek (Ed.), *Teacher expectations.* Hillsdale, NJ: Erlbaum.

Barron, F. (1968). *Creativity and personal freedom.* Princeton, NJ: Van Nostrand.

Barron, F., & Harrington, D. M. (1981). Creativity, intelligence, and personality. *Annual Review of Psychology, 32,* 439–476.

Barron, F., & Welsh, G. S. (1952). Artistic perception as a factor in personality style: Its measurement by a figure preference test. *Journal of Psychology, 33,* 199–203.

Barsalou, L. W. (1987). The instability of graded structure: Implications for the nature of concepts. In U. Neisser (Ed.), *Concepts and conceptual development: Ecological and intellectual factors in categorization.* New York: Cambridge University Press.

Barstis, S. W., & Ford, L. H., Jr. (1977). Reflection-impulsivity, conservation and the development of ability to control cognitive tempo. *Child Development, 48,* 653–659.

Bar-Tal, D., & Bar-Zohar, Y. (1977). The relationship between perception of locus of control and academic achievement. *Contemporary Educational Psychology, 2,* 181–199.

Bartlett, F. C. (1932). *Remembering.* New York: Cambridge University Press.

Battle, E. S., & Rotter, J. B. (1963). Children's feelings of personal control as related to social class and ethnic group. *Journal of Personality, 31,* 482–490.

Benbow, C. P., & Stanley, J. (1980). Sex differences in mathematical ability: Fact or artifact? *Science, 210,* 1262–1264.

Benbow, C. P., & Stanley, J. (1987). Sex differences in mathematical reasoning ability: More facts. In M. R. Walsh (Ed.), *The psychology of women: Ongoing debates.* New Haven, CT: Yale University Press.

Berglas, S. (1987). Self-handicapping model. In H. T. Blane & K. E. Leonard (Eds.), *Psychological theories of drinking and alcoholism.* New York: Guilford.

Berglas, S., & Jones, E. E. (1978). Drug choice as a self-handicapping strategy in response to noncontingent success. *Journal of Personality and Social Psychology, 36,* 403–417.

Berlin, B., & Kay, P. (1969). *Basic color terms: Their universality and evolution.* Berkeley, CA: University of California Press.

Berlyne, D. E. (1950). Novelty and curiosity as determinants of exploratory behavior. *British Journal of Psychology, 41,* 68–80.

Berlyne, D. E. (1966). Exploration and curiosity. *Science, 153,* 25–33.

Berry, J. W. (1974). Radical cultural relativism and the concept of intelligence. In J. W. Berry & P. R. Dasen (Eds.), *Culture and cognition: Readings in cross-cultural psychology.* London: Methuen.

Berry, J. W. (1976). *Human ecology and cognitive style.* New York: Wiley.

Berry, J. W. (1981). Cultural universality of any theory of human intelligence remains an open question. *Behavioral & Brain Science, 3,* 584–585.

Berry, J. W. (1982). Culture systems and cognitive styles. In M. Friedman, J. P. Das & N. O'Connor (Eds.), *Intelligence and learning.* New York: Plenum.

Berry, J. W., & Irvine, S. H. (1986). Bricolage: Savages do it daily. In R. J. Sternberg & R. K. Wagner (Eds.), *Practical intelligence: Nature and origins of competence in the everyday world.* New York: Cambridge University Press.

Bieri, J. (1955). Cognitive complexity-simplicity and predicted behavior. *Journal of Abnormal and Social Psychology, 51,* 263–268.

Bjorklund, D. F. (1988). Acquiring a mnemonic: Age and category knowledge effects. *Journal of Experimental Child Psychology, 45,* 71–87.

Bjorklund, D. F. (1989). *Children's thinking: Developmental function and individual differences.* Pacific Grove, CA: Brooks/Cole.

Bjorklund, D. F., & Muir, J. (1989). Knowledge, mental effort, and memory strategies: Developmental and individual differences. In W. Schneider & F. Weinert (Eds.), *Interactions among aptitude, strategies, and knowledge in cognitive performance.* New York: Springer-Verlag.

Block, N. J., & Dworkin, G. (1974). IQ: Heritability and inequality, Part I. *Philosophy & Public Affairs, 3,* 331–409.

Bloom, A. H. (1981). *The linguistic shaping of thought: A study in the impact of language on thinking in China and the West.* Hillsdale, NJ: Erlbaum.

Bock, W. E. (1969). Farmer's daughter effect: The case of the negro female professionals. *Phylon, 30,* 17–26.

Bogue, D. (1963). *Skid row in American cities.* Chicago, IL: University of Chicago Press.

Bolles, E. B. (1988). *Remembering and forgetting: An inquiry into the nature of memory.* New York: Walker.

Bookman, M. J. (1990). *Sources of competence in resilient children: Implications for the study of children of alcoholics.* Unpublished paper, Emory University, Atlanta, GA.

Boring, E. G. (1923, June). Intelligence as the tests test it. *New Republic,* 35–37.

Boring, E. G. (1950). *History of experimental psychology* (2nd ed.). New York: Appleton-Century-Crofts.

Borkowski, J. G., Peck, V. A., Reid, M. K., & Kurtz, B. E. (1983). Impulsivity and strategy transfer: Metacognition as a mediator. *Child Development, 54,* 459–473.

Borkowski, J. G., Reid, M. K., & Kurtz, B. E. (1984). Metacognition and retardation: Paradigmatic, theoretical, and applied perspectives. In R. Sperber, C. M.

Cauley, & P. Brooks (Eds.), *Learning and cognition in the mentally retarded.* Baltimore, MD: University Park Press.

Borkowski, J. G., & Turner, L. A. (1990). Transituational characteristics of metacognition. In W. Schneider & F. E. Weinert (Eds.), *Interactions among aptitudes, strategies, and knowledge in cognitive performance.* New York: Springer-Verlag.

Bourdieu, P. (1977). Cultural reproduction and social reproduction. In J. Karabel & A. H. Halsey (Eds.), *Power and ideology in education.* New York: Oxford University Press.

Bower, G. H. (1981). Mood and memory. *Psychology Today,* 60–69.

Bowles, S. (1977). Unequal education and the reproduction of the social division of labor. In J. Karabel & A. H. Halsey (Eds.), *Power and ideology in education.* New York: Oxford University Press.

Boyer, C. B. (1985). *A history of mathematics.* Princeton, NJ: Princeton University Press.

Boykin, A. W. (1986). The triple quandary and the schooling of Afro-American children. In U. Neisser (Ed.), *The school achievement of minority children: New perspectives.* Hillsdale, NJ: Erlbaum.

Braddock, J. H., & McPartland, J. M. (1987). How minorities continue to be excluded from equal employment opportunities: Research on labor market and institutional barriers. *Journal of Social Issues, 43,* 5–29.

Brainerd, C. J., Kingma, J., & Howe, M. I. (1986). Spread of encoding and the development of organization in memory. *Canadian Journal of Psychology, 40,* 203–223.

Brand, E. S., Ruiz, R. A., & Padilla, A. M. (1974). Ethnic identification and preference: A review. *Psychological Bulletin, 81,* 860–890.

Brenner, M. H. (1984). *Estimating the effects of economic change on national health and social well-being.* Washington, DC: U.S. Government Printing Office.

Brewin, C. R. (1985). Depression and causal attributions: What is their relation? *Psychological Bulletin, 98,* 247–309.

Brodt, S. E., & Zimbardo, P. G. (1981). Generating reward and cost orientation through praise and criticism. *Journal of Personality and Social Psychology, 35,* 437–449.

Brody, N. (1963). N achievement, test anxiety, and subjective probability of success in risk taking behavior. *Journal of Abnormal and Social Psychology, 66,* 413–418.

Bronfenbrenner, U. (1970). *Two worlds of childhood: U.S. and U.S.S.R.* New York: Simon & Schuster.

Bronfenbrenner, U. (1975). Nature with nurture: A reinterpretation of the evidence. In A. Montague (Ed.), *Race and IQ.* New York: Oxford University Press.

Bronfenbrenner, U. (1986). Ecology of the family as a context for human development: Research perspectives. *Development Psychology, 22,* 723–742.

Brophy, J., & Evertson, C. M. (1981). *Student characteristics and teaching.* New York: Longman Press.

Broverman, I. K., Vogel, S. R., Broverman, D. M., Clarkson, F. E., & Rosenkrantz, P. S. (1972). Sex-role stereotypes: A current appraisal. *Journal of Social Issues, 28,* 59–78.

Brown, A. L. (1975). The development of memory: Knowing about knowing, and knowing how to know. In H. W. Reese (Ed.), *Advances in child development and behavior* (Vol. 10). New York: Academic.

Brown, A. L., & Campione, J. C. (1984). Three faces of transfer: Implications for early competence, individual differences, and instruction. In M. E. Lamb, A. L. Brown & B. Rogoff (Eds.), *Advances in developmental psychology* (Vol. III). Hillsdale, NJ: Erlbaum.

Brown, A. L., Palincsar, A. S., & Purcell, L. (1986). Poor readers: Teach, don't label. In U. Neisser (Ed.), *The school achievement of minority children.* Hillsdale, NJ: Erlbaum.

Brown, A. L., & Reeve, H. (1987). Bandwidths of competence: The role of supportive contexts in learning and development. In L. S. Liben (Ed.), *Development and learning: Conflict and congruence.* Hillsdale, NJ: Erlbaum.

Brown, R. (1965). *Social psychology.* New York: Free Press.

Brown, R. (1973). *A first language.* Cambridge, MA: Harvard University Press.

Brown, R. (1986). *Social psychology* (2nd ed.). New York: Free Press.

Brown, R., & Hanlon, C. C. (1970). Derivational complexity and order of acquisition in child speech. In J. R. Hayes (Ed.), *Cognition and the development of language.* New York: Wiley.

Brown, R., & Lenneberg, E. (1954). A study in language and cognition. *Journal of Abnormal and Social Psychology, 49,* 454–462.

Bruner, J. (1983). *Children's talk: Learning to use language.* New York: Norton.

Bruner, J. S. (1985). *Child's talk: Learning to use language.* New York: Norton.

Bryant, B. K., & Trockel, J. F. (1976). Personal history of psychological stress related to locus of control orientation among college women. *Journal of Consulting and Clinical Psychology, 44,* 266–271.

Buie, J. (1988). "Me" decade generates depression. *The APA Monitor, 19,* 18.

Bureau of Census. (1987). *Statistical abstract of the United States* (107th ed.). Washington, DC: U.S. Department of Commerce.

Bureau of Census. (1990). *Statistical abstract of the United States* (110th ed.). Washington, DC: U.S. Department of Commerce.

Burt, C. (1966). The genetic determination of differences in intelligence: A study of monozygotic twins reared together and apart. *British Journal of Psychology, 57,* 137–153.

Campbell, J. I. D. (1987). The role of associative interference in learning and retrieving arithmetic facts. In J. A. Sloboda & D. Rogers (Eds.), *Cognitive processes in mathematics.* Oxford: Clarenon.

Campbell, J. I. D., & Graham, D. J. (1985). Mental multiplication skills: Structure, process and acquisition. *Canadian Journal of Psychology, 39,* 338–366.

Campbell, V. N. (1960). *Assumed similarity, perceived sociometric balance, and social influence.* Unpublished doctoral dissertation, University of Colorado, Boulder.

Cantor, N., & Mischel, W. (1977). Traits as prototypes: Effects on recognition memory. *Journal of Personality and Social Psychology, 35,* 38–48.

Carpenter, E. S. (1955). Space concepts of the Aivilik Eskimo. *Explorations, 5,* 131–145.

Carr, M., Kurtz, B. E., Schneider, W., Turner, L. A., & Borkowski, J. G. (1989). Strategy acquisition and transfer: Environmental influences on metacognitive development. *Developmental Psychology, 25,* 765–771.

Carroll, J. B., & Casagrande, J. (1958). The function of language classification in behavior. In E. E. Maccoby, T. M. Newcomb, & E. L. Harley (Eds.), *Readings in social psychology* (3rd ed.). New York: Holt, Rinehart and Winston.

Case, R. (1985). *Intellectual development: Birth to adulthood.* New York: Academic.

Cavanaugh, J. C., & Perlmutter, M. (1982). Metamemory: A critical examination. *Child Development, 53,* 11–28.

Ceci, S. J. (1980). A developmental study of multiple encoding and its relationship to age-related changes in free recall. *Child Development, 51,* 892–895.

Ceci, S. J. (1990). *On intelligence . . . more or less: A bio-ecological treatise on intellectual development.* Englewood Cliffs, NJ: Prentice-Hall.

Ceci, S. J., & Liker, J. (1986). A day at the races: A study of IQ, expertise, and cognitive complexity. *Journal of Experimental Psychology: General, 115,* 255–266.

Ceci, S. J., & Nightingale, N. N. (1990). The entanglement of knowledge and process in development: Toward a tentative framework for understanding individual differences in intellectual development. In W. Schneider & F. E. Weinert (Eds.), *Interactions among aptitudes, strategies, and knowledge in cognitive performance.* New York: Springer-Verlag.

Charlesworth, W. R. A. (1976). Human intelligence as adaptation: An ecological approach. In L. B. Resnick (Ed.), *The nature of intelligence.* Hillsdale, NJ: Erlbaum.

Charlesworth, W. R. A. (1979). An ecological approach to studying intelligence. *Human Development, 22,* 212–216.

Charness, N. (1983). Age, skill, and bridge bidding: A chronometric analysis. *Journal of Verbal Learning and Verbal Behavior, 22,* 406–416.

Chase, W. G., & Simon, H. A. (1973). Perception in chess. *Cognitive Psychology, 4,* 55–81.

Cheney, A. B., & Bleker, E. G. (1982). *Internal-external locus of control and repression sensitization in battered women.* Paper presented at the APA Convention, Washington, DC.

Cherry, E. C. (1953). Some experiments on the recognition of speech, with one and two ears. *Journal of the Acoustical Society of America, 25,* 975–979.

Chi, M. (1978). Knowledge structure and memory development. In R. S. Siegler (Ed.), *Children's thinking: What develops?* Hillsdale, NJ: Erlbaum.

Chi, M. T. H. (1984). Representing knowledge and meta-knowledge: Implications for interpreting metamemory research. In R. H. Kluwe & F. E. Weinert (Eds.), *Metacognition, motivation, and learning.* Hillsdale, NJ: Erlbaum.

Chi, M. T. H. (1985). Changing conception of sources of memory development. *Human Development, 28,* 50–56.

Chi, M. T. H., & Ceci, S. J. (1987). Content knowledge: Its restructuring with memory development. In H. W. Reeve & R. Lipsett (Eds.), *Advances in Child Development and Behavior, 20,* 91–146.

Chi, M. T. H., & Glaser, R. (1985). Problem-solving ability. In R. J. Sternberg (Ed.), *Human abilities: An information-processing approach.* New York: Freeman.

Child, I. L., Storm, T., & Veroff, J. (1958). Achievement themes in folktales related to socialization practices. In J. W. Atkinson (Ed.), *Motives in fantasy, action, and society.* Princeton, NJ: Van Nostrand.

Chomsky, N. (1965). *Aspects of the theory of syntax.* Cambridge, MA: MIT Press.

Chomsky, N. (1975). *Reflections on language.* New York: Pantheon.

Chomsky, N. (1980). *Rules and representations.* New York: Columbia University Press.

Chomsky, N. (1986). *Knowledge of language: Its nature, origin, and use.* New York: Praeger.

Clark, E. V. (1973). What's in a word? On the child's acquisition of semantics in his first language. In T. E. Moore (Ed.), *Cognitive development and the acquisition of language.* New York: Academic.

Clark, K. B. (1980). Kenneth B. Clark, In R. Evans (Interviewer), *The making of social psychology: Discussions with creative contributors.* New York: Gardner.

Clark, K. B., & Clark, M. P. (1939). The development of consciousness of self and the emergence of racial identity in Negro preschool children. *Journal of Social Psychology, 10,* 591–599.

Clark, K. B., & Clark, M. P. (1947). Racial identification and racial preference in Negro children. In T. M. Newcomb & E. L. Hartley (Eds.), *Readings in social psychology.* New York: Holt.

Clark, R. (1983). *Family life and school achievement: Why poor black children succeed or fail.* Chicago, IL: University of Chicago Press.

Cohen, A. R. (1959). Some implications of self-esteem for social influence. In C. I. Hovland & I. L. Janis (Eds.), *Personality and persuasibility.* New Haven, CT: Yale University Press.

Cohen, L. B., & Strauss, M. S. (1979). Concept acquisition in the human infant. *Child Development, 50,* 419–424.

Cole, M., Gay, J., Glick, J. D., & Sharp, D. W. (1971). *The cultural context of learning and thinking.* New York: Basic Books.

Cole, M., & Means, B. (1981). *Comparative studies of how people think: An introduction.* Cambridge, MA: Harvard University Press.

Cole, M., & Scribner, S. (1973). Cognitive consequences of formal and informal education. *Science, 182,* 553–559.

Cole, M., & Scribner, S. (1974). *Culture and thought: A psychological introduction.* New York: Wiley.

Coleman, J. S., Campbell, E. Q., Hobson, C. J., McPartland, J., Mood, A. M., Weinfeld, F. D., & York, R. L. (1966). *Equality of educational opportunity.* Washington, DC: U.S. Government Printing Office.

The College Board. (1985). *Equality and excellence: The educational status of black Americans.* New York: College Board.

Collier, G. (1985). *Emotional expression.* Hillsdale, NJ: Erlbaum.

Collier, G., Minton, H. L., & Reynolds, G. (1991). *Currents of thought in American social psychology.* New York: Oxford University Press.

Collins, B. E. (1974). Four separate components of the Rotter I-E Scale: Belief in a difficult world, a just world, a predictable world and a political world. *Journal of Personality and Social Psychology, 41,* 471–492.

Collins, J. (1986). Differential instructions in reading groups. In J. Cook-Gumperz (Ed.), *The social construction of literacy.* New York: Cambridge University Press.

Comer, J. P. (1989). The school development program: A psychosocial model of school intervention. In G. L. Berry & J. K. Asamen (Eds.), *Black students: Psychological issues and academic achievement.* Newbury Park, CA: Sage.

Condry, J. (1977). Enemies of exploration: Self-initiated versus other-initiated learning. *Journal of Personality and Social Psychology, 35,* 459–477.

Consortium for Longitudinal Studies. (1983). *As the twig is bent . . . lasting effects of preschool programs.* Hillsdale, NJ: Erlbaum.

Cook, S. W. (1979). Social science and school desegregation: "Did we mislead the Supreme Court?" *Personality and Social Psychological Bulletin, 5,* 420–437.

Corteen, R., & Williams, T. (1986). Television and reading skills. In T. M. Williams (Ed.), *The impact of television: A natural experiment in three communities.* New York: Academic.

Cowan, W. M. (1979). The development of the brain. *Scientific American, 241,* 112–133.

Coyne, J. C., & Gotlib, I. H. (1983). The role of cognition in depression: A critical appraisal. *Psychological Bulletin, 94,* 472-505.

Crandall, V. C. (1969). Sex differences in expectancy of intellectual and academic reinforcement. In C. P. Smith (Ed.), *Achievement-related motives in children.* New York: Russell Sage Foundation.

Crandall, V. C. (1973). *Differences in parental antecedents of internal-external control in children and in young adults.* Paper presented at the APA Convention, Montreal.

Crandall, V. C., Katkovsky, W., & Crandall, V. J. (1965). Children's belief in their own control of reinforcement in intellectual-academic achievement situations. *Child Development, 36,* 91–109.

Crawford, J. (1989). *Bilingual education: History, politics, theory and practice.* Trenton, NJ: Crane.

Crockenberg, S. B. (1984). Support for adolescent mothers during the postnatal period: Theory and research. In Z. Boukvois (Ed.), *Research on support for parents and infants in the postnatal period.* Norwood, NJ: Ablex.

Crockett, W. H. (1965). Cognitive complexity and impression formation. In B. A. Maher (Ed.), *Process in experimental personality research* (Vol. 2). New York: Academic.

Cunningham, W. R., & Owens, W. A. (1983). The Iowa State study of the adult development of intellectual abilities. In K. W. Schaif (Ed.), *Longitudinal studies of adult psychological development.* New York: Guilford.

Dacey, J. S. (1989). *Fundamentals of creative thinking.* Lexington, MA: Lexington Books.

Dansky, J. L. (1980). Make-believe: A mediator of the relationship between play and associative fluency. *Child Development, 51,* 576–579.

Darlington, R. B. (1986). Long-term effects of preschool programs. In U. Neisser (Ed.), *The school achievement of minority children: New perspectives.* Hillsdale, NJ: Erlbaum.

Darwin, C. (1859). *Origin of species.* New York: Hurst.

Davis, W. L., & Phares, E. J. (1967). Internal-external control as a determinant of information seeking in a social influence situation. *Journal of Personality, 35,* 547–561.

Deaux, K. (1976). Sex: A perspective on the attribution process. In J. H. Harvey, W. J. Ickes, & R. F. Kidd (Eds.), *New directions in attribution research.* Hillsdale, NJ: Erlbaum.

Deaux, K., & Emswiller, T. (1974). Explanation of successful performance on sex-linked tasks: What is skill for males is luck for females. *Journal of Personality and Social Psychology, 29,* 80–85.

deCharms, R. (1968). *Personal causation: The internal affective determinants of behavior.* New York: Academic.

deCharms, R. (1981). Personal causation and locus of control: Two different traditions and two uncorrelated measures. In H. M. Lefcourt (Ed.), *Research with the locus of control construct* (Vol. 1). New York: Academic.

Deci, E. L. (1971). The effects of externally mediated rewards on intrinsic motivation. *Journal of Personality and Social Psychology, 18,* 105–115.

Deci, E. L. (1972). Intrinsic motivation, extrinsic reinforcement, and inequity. *Journal of Personality and Social Psychology, 22,* 113–120.

Deci, E. L. (1975). *Intrinsic motivation.* New York: Plenum.

Deci, E. L. (1980). *The psychology of self-determination.* Lexington, MA: Lexington Books.

Deci, E. L., Cascio, W. F., & Krusell, J. (1975). Cognitive evaluation theory and some comments on the Calder and Shaw critique. *Journal of Personality and Social Psychology, 31,* 81–85.

de Grout, A. D. (1965). *Thought and choice in chess.* The Hague: Mouton.

Dembo, T. (1931). Der Anger als dynamisches Problem [Anger as a dynamic problem]. *Psychologische Forschung, 15,* 1–144.

Dennis, W. (1966). Creative productivity between the ages of 20 and 80 years. *Journal of Gerontology, 21,* 1–8.

de Villers, J. G., & de Villers, P. A. (1979). *Early language.* Cambridge, MA: Harvard University Press.

DeVos, G. (1965). Achievement orientation, social self-identity and Japanese economic development. *Asian Survey, 5,* 575–589.

Dewey, J. (1922). *Human nature and conduct: An introduction to social psychology.* New York: Holt.

Dewey, J. (1933). *How we think: A restatement of the relation of reflective thinking to the educative process* (rev. ed.). Boston, MA: Heath. (Original work published 1911)

Dewey, J. (1957). *Human nature and conduct.* New York: Modern Library.

Diener, C., & Dweck, C. (1978). An analysis of learned helplessness: Continuous changes in performance, strategy, and achievement cognitions following failure. *Journal of Personality and Social Psychology, 36,* 351–362.

Doherty, W. J. (1983). Impact of divorce on locus of control orientation in adult women: A longitudinal study. *Journal of Personality and Social Psychology, 44,* 834–840.

Dorner, D., & Kreuzig, H. (1983). Problemlosefahigkeit und intelligenz. *Psychologischs Rundschaus, 34,* 185–192.

Douvan, E. (1958). Social status and success striving. In J. W. Atkinson (Ed.), *Motives in fantasy, action, and society.* Princeton, NJ: Van Nostrand.

Dreyfus, H. L., & Dreyfus, S. E. (1988). Making a mind versus modeling the brain: Artificial intelligence back at a branchpoint. In S. R. Graubard (Ed.), *The artificial intelligence debate: False starts, real foundations.* Cambridge, MA: MIT Press.

Dubé, E. F. (1982). Literacy, culture familiarity, and "intelligence" as determinants of story recall. In U. Neisser (Ed.), *Memory observed: Remembering in natural context.* San Francisco, CA: Freeman.

DuBois, W. E. B. (1903). *Souls of black folk.* Chicago, IL: McClurg.

Duffy, G. G., Roehler, L. R., Book, C., Meloth, M. S., & Vaurus, L. G. (1984). *Instructional characteristics which promote strategic awareness in reading.* Paper presented at the Annual meeting of the American Educational Research Association, New Orleans.

Duke, M. P., & Lancaster, W. A. (1976). A note on locus of control as a function of father absence. *Journal of Genetic Psychology, 127,* 335–336.

Duke, M. P., & Lewis, G. (1979). The measurement of locus of control on black preschool and primary school. *Journal of Personality Assessment, 43,* 351–355.

Duke, M. P., & Nowicki, S. (1974). Locus of control and achievement—conformation of a theoretical expectation. *Journal of Psychology, 87,* 263–267.

Duncker, K. (1945). On problem solving. *Psychological Monograph, 58* (5, Whole No. 270).

Dunkin, D. (1978–79). What classroom observation reveals about reading comprehension instruction. *Reading Research Quarterly, 14,* 481–533.

Dusek, J. B. (Ed.). (1985). *Teachers expectancies.* Hillsdale, NJ: Erlbaum.

Dweck, C. S. (1975). The role of expectations and attributions in the alleviation of learned helplessness. *Journal of Personality and Social Psychology, 31,* 674–685.

Dweck, C. S. (1986). Motivational processes affecting learning. *American Psychologist, 41,* 1040–1048.

Dweck, C. S., & Bush, E. S. (1976). Sex differences in learned helplessness: I. Differential debilitation with peer and adult evaluators. *Developmental Psychology, 12,* 147–156.

Dweck, C. S., & Elliot, E. S. (1983). Achievement motivation. In P. H. Mussen (Ed.), *Handbook of child psychology.* New York: Wiley.

Dweck, C. S., Goetz, T. E., & Strauss, N. (1980). Sex differences in learned helplessness: IV. An experimental and naturalistic study of failure generalization and its mediators. *Journal of Personality and Social Psychology, 38,* 441-452.

Dweck, C. S., & Leggett, E. L. (1988). A social-cognitive approach to motivation and personality. *Psychological Review, 95,* 256–273.

Dweck, C. S., & Licht, B. G. (1980). Learned helplessness and academic achievement. In J. Garber & M. Seligman (Eds.), *Human helplessness: Theory and application.* New York: Academic.

Dweck, C. S., & Reppucci, N. D. (1973). Learned helplessness and reinforcement responsibility in children. *Journal of Personality and Social Psychology, 25,* 109–116.

Eccles, J. S., & Jacobs, J. E. (1987). Social forces shape math aptitude and performance. In M. R. Walsh (Ed.), *The psychology of women: Ongoing debates.* New Haven, CT: Yale University Press.

Edmonds, R. (1979). Effective schools for the urban poor. *Educational Leadership, 37,* 15–23.

Edmonds, R. (1986). Characteristics of effective schools. In U. Neisser (Ed.), *The school achievement of minority children: New perspectives.* Hillsdale, NJ: Erlbaum.

Egan, D. E., & Schwartz, B. J. (1979). Chunking in recall of symbolic drawing. *Memory and Cognition, 7,* 149–158.

Eimas, P. D., Siqueland, E. R., Jusczyk, P., & Vigorito, J. (1971). Speech perception in infants. *Science, 171,* 303–306.

Eisenstadt, J. M. (1978). Parental loss and genius. *American Psychologist, 33,* 211–223.

Elliot, E., & Dweck, C. S. (1985). Goals: An approach to motivation and achievement. *Journal of Personality and Social Psychology, 54,* 5–12.

Ellis, L. J., & Bentler, P. M. (1973). Traditional sex-determinants: Role standards and sex stereotypes. *Journal of Personality and Social Psychology, 24,* 28–34.

Ellis, N. C., & Hennelley, R. A. (1980). A bilingual word-length effect: Implications for intelligence testing and the relative ease of mental calculation in Welsh and English. *British Journal of Psychology, 71,* 43–52.

Epps, E. G. (1975). Impact of school desegregation on aspirations, self-concepts and other aspects of personality. *Law and Contemporary Problems, 39,* 300–313.

Epstein, A. (1980). *Assessing the child development information needed by adolescent parents with very young children.* Final report of Grant OCD-90-C-1341. Washington, DC: Department of Health, Education and Welfare, Office of Child Development.

Epstein, S. (1985). The implications of cognitive-experiential self-theory for research in social psychology and personality. *Journal for the Theory of Social Behaviour, 15,* 283–310.

Epstein, S., & Meier, P. (1989). Constructive thinking: A broad coping variable with specific components. *Journal of Personality and Social Psychology, 57,* 332–350.

Erikson, E. H. (1963). *Childhood and society.* New York: Norton.

Eysenck, H. J. (1982). *A model for intelligence.* New York: Springer-Verlag.

Fantz, R. L. (1961). The origin of form perception. *Scientific American, 206,* 66–72.

Feather, N. T. (1961). The relationship of persistence at a task to expectation of success and achievement related motive. *Journal of Abnormal and Social Psychology, 63,* 552–561.

Feather, N. T. (1963). Persistence at a difficult task with alternative task of intermediate difficulty. *Journal of Abnormal and Social Psychology, 66,* 604–609.

Feather, N. T. (1965). The relationship of expectation of success to need achievement and test anxiety. *Journal of Personality and Social Psychology, 1,* 118–126.

Feather, N. T., & Raphelson, A. C. (1974). Fear of success in Australian and American student groups: Motive or sex-role stereotype. *Journal of Personality, 42,* 190–201.

Feather, N. T., & Simon, J. G. (1975). Reactions to male and female success and failure in sex-linked occupations: Impressions of personality, causal attributions, and perceived likelihood of different consequences. *Journal of Personality and Social Psychology, 31,* 20–31.

Feldman-Summers, S., & Kiesler, S. B. (1974). Those who are number two try harder: The effect of sex on attributions of causality. *Journal of Personality and Social Psychology, 30,* 846–854.

Festinger, L. (1950). Informal social communication. *Psychological Review, 57,* 271–282.

Festinger, L. (1951). Informal communication in small groups. In H. Gurtzkow (Ed.), *Groups, leadership, and men.* Pittsburgh, PA: Carnegie Press.

Festinger, L. (1954). A theory of social comparison processes. *Human Relations, 7,* 117–140.

Findley, M. J., & Cooper, H. M. (1983). Locus of control and academic achievement: A literature review. *Journal of Personality and Social Psychology, 44,* 419–427.

Fingerette, H. (1969). *Self-deception.* New York: Routledge & Kegan Paul.

Flavell, J. H. (1963). *The developmental psychology of Jean Piaget.* Princeton, NJ: Van Nostrand.

Flavell, J. H. (1970). Developmental studies of mediated memory. In H. W. Reese & L. P. Lipsitt (Eds.), *Advances in child development and behavior* (Vol. 5). New York: Academic.

Flavell, J. H. (1971). First discussant's comment: What is memory development the development of? *Human Development, 14,* 272–278.

Flavell, J. H. (1978). Comments. In R. S. Siegler (Ed.), *Children's thinking: What develops?* Hillsdale, NJ: Erlbaum.

Flavell, J. H., Beach, D. H., & Chinsky, J. M. (1966). Spontaneous verbal rehearsal in memory task as a function of age. *Child Development, 37,* 283–299.

Flavell, J. H., Friedrichs, A. C., & Hoyt, J. D. (1970). Developmental changes in memorization processes. *Cognitive Psychology, 1,* 324–340.

Flynn, J. R. (1984). The mean IQ of Americans: Massive gains 1932 to 1978. *Psychological Bulletin, 95,* 29–51.

Flynn, J. R. (1987). Race and IQ: Jensen's case refuted. In S. Modgil & C. Modgil (Eds.), *Arthur Jensen: Consensus and controversy.* New York: Falmer Press.

Frankel, A., & Snyder, M. L. (1978). Poor performance following unsolvable problems: Learned helplessness or egotism. *Journal of Social Psychology, 36,* 1415–1423.

Freud, S. (1953). The interpretation of dreams. In J. Stachey (Ed.), *The standard edition of the complete psychological works of Sigmund Freud.* London: Hogarth Press. (Original work published 1900)

Freud, S. (1957). Instincts and their vicissitudes. In J. Stachey (Ed.), *The standard edition of the complete psychological works of Sigmund Freud.* London: Hogarth Press. (Original work published 1915)

Friedan, B. (1963). *The feminine mystique.* New York: Norton.

Friedrich, L. K., & Stein, A. H. (1973). Aggressive and prosocial television programs and the natural behavior of preschool children. *Monographs of the Society for Research in Child Development, 38,* 1–64.

Frieze, I. H. (1975). Women's expectation for causal attributions of success and failure. In M. T. S. Mednick, S. S. Tangri, & L. H. Hoffman (Eds.), *Women: Social psychological perspectives on achievement.* New York: Holt, Rinehart and Winston.

Furstenberg, F., & Crawford, A. (1978). Family support: Helping teenage mothers to cope. *Family Planning Perspectives, 10,* 322–333.

Gadberry, S. (1980). Effects of restricting first graders' TV viewing on leisure time use, IQ change, and cognitive style. *Journal of Applied Developmental Psychology, 1,* 45–47.

Galbraith, J. K. (1977). *The age of uncertainty.* Boston, MA: Houghton Mifflin.

Galotti, K. M. (1989). Approaches to studying formal and everyday reasoning. *Psychological Bulletin, 103,* 331–351.

Gardner, H. (1983). *Frames of mind: The theory of multiple intelligence.* New York: Basic Books.

Gardner, H. (1985). *The mind's new science: A history of the cognitive revolution.* New York: Basic Books.

Garfield, E. (1978). The one hundred articles most cited by social scientists, 1969–1977. *Current Contents, 32,* 5–14.

Garmezy, N. (1971). Vulnerability research and the issue of primary prevention. *American Journal of Orthopsychiatry, 41,* 101–116.

Gay, J., & Cole, M. (1967). *The new mathematics and an old culture.* New York: Holt, Rinehart and Winston.

Gelman, D., Springen, K., Brailsford, K., & Miller, M. (1988, March). Black and white in America. *Newsweek,* 18–23.

Gelman, R., & Brown, A. L. (1986). Changing views of cognitive competence in the young. In N. J. Smelser & D. R. Gerstein (Eds.), *Behavioral and social science: Fifty years of discovery.* Washington, DC: National University Press.

Gerard, H. B. (1983). School desegregation: The social science role. *American Psychologist, 38,* 869–877.

Getzels, J. W. (1975). Creativity: Prospects and issues. In I. A. Taylor & J. W. Getzels (Eds.), *Perspectives in creativity.* Chicago, IL: Aldine.

Gibson, E. J. (1969). *Principles of perceptual learning and development.* New York: Appleton-Century-Crofts.

Gibson, E. J. (1988). Exploratory behavior in the development of perceiving, acting, and the acquiring of knowledge. *Annual Review of Psychology, 39,* 1–41.

Gibson, E. J., & Spelke, E. S. (1983). Development of perception. In P. H. Mussen (Ed.), *Handbook of child psychology* (Vol. 3). New York: Wiley.

Gibson, J. J. (1966). *The senses considered as perceptual systems.* Boston, MA: Houghton Mifflin.

Gibson, J. J. (1979). *An ecological approach to visual perception.* Boston, MA: Houghton Mifflin.

Giddens, A. (1971). *Capitalism and modern social theory.* London: Cambridge University Press.

Gilmore, P. (1985). "Gimme room": School resistance, attitude, and access to literacy. *Journal of Education, 167,* 111–128.

Ginsburg, H. P. (1977). *Children's arithmetic: The learning process.* New York: Van Nostrand.

Ginsburg, H. P. (Ed.). (1983). *The development of mathematical thinking.* New York: Academic.

Giovannoni, J., & Billingsley, A. (1970). Child neglect among the poor: A study of parental adequacy in families of other ethnic groups. *Child Welfare, 49,* 196–204.

Gladwin, T. (1970). *East is a big bird: Navigation and logic on Puluwat atoll.* Cambridge, MA: Harvard University Press.

Glaser, R. (1984). Education and thinking: The role of knowledge. *American Psychologist, 39,* 93–104.

Glaser, R. (1987). Thoughts on expertise. In C. Schooler & K. W. Schaie (Eds.), *Cognitive function and social structure over the life course.* Norwood, NJ: Ablex.

Glaser, R. (1989). *The fourth R, the ability to reason.* Washington, DC: Federation of Behavioral, Psychological and Cognitive Sciences.

Glass, D. C., & Singer, J. E. (1972). *Urban stress: Experiments on noise and social stressors.* New York: Academic.

Glass, D. C., Singer, J. E., Leonard, H. S., Krantz, D., Cohen, S., & Cummings, H. (1973). Perceived control of aversive stimulation and the reduction of stress responses. *Journal of Personality, 41,* 577–595.

Goddard, H. H. (1917). Mental tests and the immigrant. *Journal of Delinquency, 2,* 243–277.

Godden, D. R., & Baddeley, A. D. (1975). Context-dependent memory in two natural environments: On land and underwater. *British Journal of Psychology, 66,* 325–371.

Goertzel, M. G., Goertzel, V., & Goertzel, T. G. (1978). *300 eminent personalities.* San Francisco, CA: Jossey-Bass.

Goldstein, K. M., & Blackman, S. (1978). *Cognitive styles.* New York: Wiley.

Goodenough, D. R. (1976). The role of individual differences in field dependence as a factor in learning and memory. *Psychological Bulletin, 83,* 675–694.

Gorman, B. S. (1968). An observation of altered locus of control following political disappointment. *Psychological Reports, 23,* 1094.

Gougis, R. A. (1986). The effects of prejudice and stress on the academic performance of black-Americans. In U. Neisser (Ed.), *The school achievement of minority children: New perspectives.* Hillsdale, NJ: Erlbaum.

Gould, S. J. (1981). *The mismeasure of man.* New York: Norton.

Graham, D. J. (1987). An associative retrieval model of arithmetic memory: How children learn to multiply. In J. A. Sloboda & D. Rogers (Eds.), *Cognitive processes in mathematics.* Oxford: Clarendon.

Graham, S. (1988). Can attribution theory tell us something about motivation in blacks? *Educational Psychology, 23,* 3–21.

Graham, S. (1989). Motivation in Afro-Americans. In G. L. Berry & J. K. Asaner (Eds.), *Black students: Psychosocial issues and academic achievement.* Newbury Park, CA: Sage.

Greenberg, H., & Fare, D. (1959). An investigation of several variables and determinants of authoritarianism. *Journal of Social Psychology, 49,* 105–111.

Greenberg, J. (1983). *Difficult goal choice as a self-handicapping strategy.* Unpublished manuscript, Ohio State University, Columbus.

Greenough, W. T. (1976). Enduring brain effects of differential experience and training. In M. R. Rosenzweig & E. L. Bennett (Eds.), *Neural mechanisms of learning and memory.* Cambridge, MA: MIT Press.

Greenough, W. T., Larson, J. R., & Withers, G. S. (1985). Effects of unilateral and bilateral training in a reaching task on dendritic branching of neurons in the rat motor-sensory forelimb cortex. *Behavioral and Neural Biology, 44,* 301–314.

Greer, B. (1987). Understanding of arithmetical operations as models of situations. In J. A. Sloboda & D. Rogers (Eds.), *Cognitive processes in mathematics.* Oxford: Clarendon.

Grevlich, W. W. (1957). A comparison of the physical growth and development of American-born and Japanese children. *American Journal of Physical Anthropology, 15,* 232–235.

Groff, P. J. (1962). A survey of basic reading group practices. *The Reading Teacher, 15,* 232–235.

Gruber, H. E., & Davis, S. N. (1988). Inching our way up Mount Olympus: The evolving-systems approach to creative thinking. In R. J. Sternberg (Ed.), *The nature of creativity: Contemporary psychological perspectives.* New York: Cambridge University Press.

Guilford, J. P. (1967). *The nature of human intelligence.* New York: McGraw-Hill.

Guilford, J. P. (1977). *Way beyond the IQ: A guide to improving intelligence and creativity.* Buffalo, NY: Creative Educational Foundations.

Guilford, J. P. (1982). Cognitive psychology's ambiguities: Some suggested remedies. *Psychological Review, 89,* 48–59.

Guilford, J. P. (1985). The structure-of-intellect model. In B. B. Wolman (Ed.), *Handbook of intelligence: Theories, measurements, and applications.* New York: Wiley.

Haack, S. (1978). *Philosophy of logic.* London: Cambridge University Press.

Hadamard, J. (1949). *The psychology of invention in the mathematical field.* Princeton, NJ: Princeton University Press.

Hakuta, K., & Garcia, E. E. (1989). Bilingualism and education. *American Psychologist, 44,* 374–379.

Halpern, D. F. (1984). *Thought and knowledge: An introduction to critical thinking.* Hillsdale, NJ: Erlbaum.

Harlow, H. F. (1953). Mice, monkeys, men, and motives. *Psychological Review, 60,* 23–32.

Harrison, L., & Williams, T. (1986). Television and cognitive development. In T. M. Williams (Ed.), *The impact of television: A national experiment in three communities.* New York: Academic.

Harvey, J. M. (1971). Locus of control shift in administrators. *Perceptual and Motor Skills, 33,* 980–982.

Hasher, L., & Zacks, R. T. (1979). Automatic and effortful processes in memory. *Journal of Experimental Psychology General, 108,* 356–388.

Hatano, G. (1982). Cognitive consequences of practice on culture specific skills. *Quarterly Newsletter of the Laboratory of Comparative Human Cognition, 4,* 15–18.

Hawkins, M. L. (1966). Mobility of students in reading groups. *The Reading Teacher, 20,* 136–140.

Hawley, P. (1971). What women think men think: Does it affect their career choice? *Journal of Consulting and Clinical Psychology, 18,* 193–199.

Hearnshaw, L. S. (1979). *Cyril Burt: Psychologist.* Ithaca, NY: Cornell University Press.

Heath, S. B. (1989). Oral and literate traditions among Black Americans living in poverty. *American Psychologist, 44,* 367–373.

Hebb, D. O., Lambert, W. E., & Tucker, G. R. (1973). A DMZ in the language war. *Psychology Today,* 55–62.

Heider, E. R., & Oliver, D. C. (1972). The structure of color space in naming and memory for two languages. *Cognitive Psychology, 3,* 337–354.

Heider, F. (1958). *The psychology of interpersonal relationships.* New York: Wiley.

Helmreich, R. L., Spence, J. T., & Gibson, R. L. (1982). Sex-role attitudes: 1972–1980. *Personality and Social Psychological Bulletin, 8,* 656–663.

Helms, J. E. (1992). Why is there no study of cultural equivalence in standardized cognitive ability testing? *American Psychologist, 47,* 1083–1101.

Henderson, N. D. (1982). Human behavior genetics. *Annual Review of Psychology, 33,* 403–440.

Hetherington, E. M. (1972). Effects of father absence on personality development in adolescent daughters. *Developmental Psychology, 7,* 313–326.

Heyns, B. (1978). *Summer learning and the effects of schooling.* New York: Academic.

Hitch, G., Cundick, J., Haughey, M., Pugh, R., & Wright, H. (1987). Aspects of counting in children's arithmetic. In J. A. Sloboda & D. Rogers (Eds.), *Cognitive processes in mathematics.* Oxford: Clarendon.

Hobson, P. (1990, May). *Only connect? Through feelings and sight to self and symbol.* Paper presented at the Conference on the Interpersonal Self, Emory University, Atlanta, GA.

Hoffman, C., Lau, I., & Johnson, D. R. (1986). The linguistic relativity of person cognition: An English-Chinese comparison. *Journal of Personality and Social Psychology, 51,* 1097–1105.

Hoffman, L. W. (1974). Fear of success in males and females: 1965 and 1972. *Journal of Consulting and Clinical Psychology, 42,* 353–358.

Honig, B. (1987). High standards and great expectations: The foundations for student achievement. In D. S. Strickland & E. J. Cooper (Eds.), *Educating black children: America's challenge.* Washington, DC: Howard University Press.

Horner, M. (1968). *Sex differences in achievement motivation and performance in competitive and noncompetitive situations.* Unpublished doctoral dissertation, University of Michigan, Ann Arbor, MI.

Horner, M. (1972). Toward an understanding of achievement-related conflict in women. *Journal of Social Issues, 28,* 157–175.

Hull, C. L. (1943). *Principles of behavior.* New York: Appleton-Century-Crofts.

Hunt, E. (1976). Varieties of cognitive power. In L. B. Resnick (Ed.), *The nature of intelligence.* Hillsdale, NJ: Erlbaum.

Hunt, E. B. (1978). Mechanics of verbal ability. *Psychological Review, 85,* 109–130.

Hurlbert, A., & Poggio, T. (1988). Making machines (and artificial intelligence) see. In S. R. Graubard (Ed.), *The artificial intelligence debate: False starts, real foundations.* Cambridge, MA: MIT Press.

Huttenlocher, P. R. (1979). Synaptic density in human frontal cortex—Developmental changes and effects of aging. *Brain Research, 163,* 193–205.

Huxley, T. H. (1935). *The diary of the voyage of H.M.S. Rattlesnake.* (J. Huxley, Ed.). London: Chatto & Windus.

Irvine, J. J. (1985). Teacher communication patterns as related to the race and sex of student. *Journal of Education Research, 78,* 338–345.

Irvine, J. J. (1986). Teacher-student interactions: Effects of student race, sex, and grade level. *Journal of Educational Psychology, 78,* 14–21.

Irvine, J. J. (1988). Urban schools that work: A summary of relevant factors. *Journal of Negro Education, 57,* 26–42.

Irvine, J. J. (1990). *Black students and school failure: Policies, practices, and prescriptions.* New York: Greenwood Press.

Irvine, R. W., & Irvine, J. J. (1983). The impact of the desegregation process on the education of black students: Key variables. *Journal of Negro Education, 52,* 410–422.

Isaac, W. (1962). Evidence for a sensory drive in monkeys. *Psychological Reports, 11,* 175–181.

Isaacson, R. L. (1964). Relation between achievement, test anxiety, and curricular choices. *Journal of Abnormal and Social Psychology, 68,* 447–452.

Jackson, B. (1964). *Streaming: An educational system in miniature.* London: Routledge & Kegan Paul.

James, W. (1890). *The principles of psychology.* New York: Holt.

Jeffries, R., Turner, A. A., Polson, P. C., & Attwood, M. E. (1981). The processes involved in designing software. In J. R. Anderson (Ed.), *Cognitive skills and their acquisition.* Hillsdale, NJ: Erlbaum.

Jencks, C., Smith, M., Acland, H., Bane, M. J., Cohen, D., Gintis, H., Heyns, B., & Michelsen, S. (1972). *Inequality: A reassessment of the effects of family and schooling in America.* New York: Basic Books.

Jensen, A. R. (1969). How much can we boost IQ and scholastic achievement? *Harvard Educational Review, 39,* 1–123.

Joe, T. (1982). *Profiles of families in poverty: Effects of the FY 1983 budget proposal on the poor.* Washington, DC: Center for the Study of Social Policy.

Johnson, C. M., Bradley-Johnson, S., McCarthy, R., & Jamie, M. (1984). Token reinforcement during WISC-R administration. *Applied Research on Mental Retardation, 5,* 43–52.

Johnson, J. E. (1976). Relations of divergent thinking and intelligence test scores with social and non-social make-believe play of preschool children. *Child Development, 47,* 1200–1203.

John-Steiner, V., & Tatter, P. (1983). An interactionist model of language development. In B. Bain (Ed.), *The sociogenesis of language and human conduct.* New York: Plenum.

Jones, E. E., & Berglas, S. (1978). Control of attributions about the self through self-handicapping strategies: The appeal of alcohol and the role of underachievement. *Personality and Social Psychology Bulletin, 4,* 200–206.

Jones, E. E., & Davis, K. E. (1965). From acts to disposition: The attribution process in person perception. In L. Berkowitz (Ed.), *Advances in experimental social psychology* (Vol. 2). New York: Academic.

Jones, J. D., Van Fossen, B. E., & Spade, J. Z. (1987, April). *Individual and organizational predictors of high school track placement.* Paper presented at the meeting of the American Educational Research Association, Washington, DC.

Julian, J. W., & Katz, S. B. (1968). Internal versus external control and the value of reinforcement. *Journal of Personality and Social Psychology, 76,* 43–48.

Kagan, J. (1965). Reflection-impulsivity and reading ability in primary grade children. *Child Development, 36,* 609–628.

Kagan, J. (1971). *Change and continuity in infancy.* New York: Wiley.

Kagan, J., Lapidus, D. R., & Moore, M. (1978). Infant antecedents of cognitive functioning: A longitudinal study. *Child Development, 49,* 1005–1023.

Kagan, J., & Moss, H. A. (1959). The stability and validity of achievement fantasy. *Journal of Abnormal and Social Psychology, 58,* 357–364.

Kagan, J., Roseman, B. L., Day, D., Albert, J., & Philips, W. (1964). Information processing in the child: Significance of analytic and reflective attitudes. *Psychological Monographs, 78* (No. 578).

Kamin, L. (1972). *Colloquium on Cyril Burt.* Department of Psychology, Princeton University, Princeton, NJ.

Kamin, L. (1974). *The science and politics of IQ.* Potomac, MD: Erlbaum.

Kantrowitz, B., & Springen, K. (1988, March). A tenuous bond from 9 to 5. *Newsweek,* 24–25.

Karabenick, S. A., & Srull, T. K. (1978). Effects of personality and situational variation in locus of control on cheating: Determinants of the "congruence effect." *Journal of Personality, 44,* 72–95.

Katkovsky, W., Crandall, V. C., & Good, S. (1967). Parental antecedents of children's beliefs in internal-external control of reinforcement in intellectual achievement situations. *Child Development, 38,* 765–776.

Kaufman, E. L., Lord, M. W., Reese, T. W., & Volkmann, J. (1949). The discrimination of visual numbers. *American Journal of Psychology, 62,* 498–525.

Kearins, J. M. (1981). Visual spatial memory in Australian Aboriginal children of desert regions. *Cognitive Psychology, 3,* 434–460.

Keating, D. P. (1984). The emperor's new clothes: The "new look" in intelligence research. In R. J. Sternberg (Ed.), *Advances in the psychology of human intelligence* (Vol. 2). Hillsdale, NJ: Erlbaum.

Keil, F. C. (1987). Conceptual development and category structure. In U. Neisser (Ed.), *Concepts and conceptual development: Ecological and intellectual factors in categorization.* New York: Cambridge University Press.

Kelley, H. H. (1967). Attribution theory in social psychology. In D. Levine (Ed.), *Nebraska symposium on motivation* (Vol. 15). Lincoln, NE: University of Nebraska Press.

Kelley, H. H. (1972). Attribution in social interaction. In E. E. Jones, D. E. Kanouse, H. H. Kelley, R. E. Nisbett, S. Valins, & B. Werner (Eds.), *Attribution: Perceiving the causes of behavior.* Morristown, NJ: General Learning Press.

Kelly, G. A. (1955). *The psychology of personal constructs.* New York: Norton.

Klahr, D., & Wallace, J. G. (1976). *Cognitive development: An information processing view.* Hillsdale, NJ: Erlbaum.

Klein, A., & Starkey, P. (1987). The origins and development of numerical cognition: A comparative analysis. In J. A. Sloboda & D. Rogers (Eds.), *Cognitive processes in mathematics.* Oxford: Clarendon.

Kline, M. (1953). *Mathematics in Western culture.* New York: Oxford University Press.

Klineberg, O. (1940). *Social psychology.* New York: Holt.

Kobasa, S. C. (1979). Stressful life events, personality, and health: An inquiry into hardiness. *Journal of Personality and Social Psychology, 37,* 1–11.

Kohlberg, L. (1966). A cognitive-developmental analysis of children's sex-role concepts and attitudes. In E. E. Maccoby (Ed.), *The development of sex differences.* Stanford, CA: Stanford University Press.

Köhler, W. (1925). *The mentality of apes.* New York: Harcourt.

Kohn, M. L. (1963). Social class and parent-child relationships: An interpretation. *American Journal of Sociology, 68,* 471–480.

Kohn, M. L. (1969). *Class and conformity: A study of values.* Homewood, IL: Dorsey.

Kohn, M. L., & Schooler, C. (1978). Reciprocal effects of the substantive complexity of work and intellectual flexibility: A longitudinal assessment. *American Journal of Sociology, 84,* 24–52.

Kolb, B. (1989). Brain development, plasticity, and behavior. *American Psychologist, 44,* 1203–1212.

Kolb, D. (1965). Achievement motivation training in underachieving high-school boys. *Journal of Personality and Social Psychology, 2,* 783–792.

Kresojevick, I. Z. (1972). Motivation to avoid success in women as related to year in school, academic achievement and success content. *Dissertation Abstract International, 33,* 2348–2349.

Kreutzer, M. A., Leonard, C., & Flavell, J. H. (1975). An interview study of children's knowledge about memory. *Monographs of the Society for Research in Child Development, 40* (Serial No. 159).

Kubie, L. S. (1958). *Neurotic distortion of the creative process.* Lawrence, KS: University of Kansas Press.

Kuhn, T. S. (1970). *The structure of scientific revolutions* (2nd ed.). Chicago, IL: University of Chicago Press.

Kurtz, B. E., & Borkowski, J. G. (1984). Children's metacognition: Exploring relations among knowledge, processes, and motivational variables. *Journal of Experimental Child Psychology, 37,* 335-354.

Kurtz, B. E., Borkowski, J. G., & Deshmuich, K. (1988). Metamemory development in Maharashtrian children: Influences from home and school. *Journal of Genetic Psychology, 149,* 363–376.

Kurtz, B. E., Schneider, W., Carr, M., Borkowski, J. G., & Turner, L. A. (1988). Sources of memory and metamemory development: Societal, parental, and educational influence. In M. Gruneverg, P. Morris, & R. Sykes (Eds.), *Proceedings of the Second International Conference on Practical Aspects of Memory* (Vol. 2). New York: Wiley.

Laboratory of Comparative Human Cognition. (1982). Culture and intelligence. In R. J. Sternberg (Ed.), *Handbook of human intelligence.* New York: Cambridge University Press.

Laboratory of Comparative Human Cognition. (1983). Culture and cognitive development. In P. Mussen (Ed.), *Handbook of child psychology.* New York: Wiley.

Langer, E. J. (1984). Playing the middle against both ends: The influence of adult cognitive activity as a model for cognitive activity in childhood and old age. In S. R. Yussen (Ed.), *The development of reflection.* New York: Academic.

Langer, E. J., & Rodin, J. (1976). The effects of choice and enhanced personal responsibility for the aged: A field experiment in an institutional setting. *Journal of Personality and Social Psychology, 34,* 191–198.

Lao, R. (1970). Internal-external control and competent and innovative behavior among Negro college students. *Journal of Personality and Social Psychology, 14,* 263–270.

Lasky, R. E., Syrdal-Lasky, A., & Klein, R. E. (1975). VOT discrimination by four to six-and-a-half-month-old infants from Spanish environments. *Journal of Experimental Child Psychology, 20,* 215–225.

Leary, M. R. (1983). *Understanding social anxiety: Social, personality, and clinical perspectives.* Beverly Hills, CA: Sage.

Leavitt, H. J., Hax, H., & Roche, J. H. (1955). "Authoritarianism" and agreement with things authoritative. *Journal of Psychology, 40,* 215–221.

Lefcourt, H. M. (1982). *Locus of control: Current trends in theory and research* (2nd ed.). Hillsdale, NJ: Erlbaum.

Lefcourt, H. M., Lewis, L. E., & Silverman, I. W. (1968). Internal versus external control of reinforcement and attention in decision making tasks. *Journal of Personality, 36,* 663–682.

Lefcourt, H. M., & Wine, J. (1969). Internal versus external control of reinforcement and the deployment of attention in experimental situations. *Canadian Journal of Behavioral Science, 1,* 127–181.

Lehman, H. C. (1953). *Age and achievement.* Princeton, NJ: Princeton University Press.

Leinhardt, G., Seewald, A., & Engel, M. (1979). Learning what's taught: Sex differences in instruction. *Journal of Educational Psychology, 71,* 432–439.

Lenski, G. E. (1961). Trends in inter-generational occupational mobility in the United States. *American Sociological Review, 23,* 514–523.

Lepper, M. R., & Greene, D. (1978). Overjustification research and beyond: Toward a means-end analysis of intrinsic and extrinsic motivation. In M. R. Lepper & D. Greene (Eds.), *The hidden costs of rewards: New perspectives on the psychology of human motivation.* Hillsdale, NJ: Erlbaum.

Lepper, M. R., Greene, D., & Nisbett, R. E. (1973). Undermining children's intrinsic interest with extrinsic reward: A test of the overjustification hypothesis. *Journal of Personality and Social Psychology, 28,* 129–137.

Levenson, H. (1973). Perceived parental antecedents of internal, powerful others, and chance locus of control orientations. *Developmental Psychology, 9,* 260–265.

Levenson, H. (1981). Differentiating among internality, powerful others, and chance. In H. M. Lefcourt (Ed.), *Research with the locus of control construct* (Vol. 1). New York: Academic.

Levin, J. R., Yussen, S. R., DeRose, T. M., & Pressley, M. (1977). Developmental changes in assessing recall and recognition memory capacity. *Developmental Psychology, 13,* 608–615.

Lewin, K. (1935). *A dynamic theory of personality.* New York: McGraw-Hill.

Lewinsohn, P. M., & Rosenbaum, M. (1987). Recall of parental behavior by acute depressives, remitted depressives and nondepressives. *Journal of Personality and Social Psychology, 52,* 611–619.

Lewontin, R. (1970). Race and intelligence. *Bulletin of the Atomic Scientists,* 2–8.

Liebert, R. M., & Morris, L. W. (1967). Cognitive and emotional components of test anxiety: A distinction and some initial data. *Psychological Reports, 20,* 975–978.

Loftus, E. F. (1979). *Eyewitness testimony.* Cambridge, MA: Harvard University Press.

Loftus, E. F., & Loftus, G. R. (1980). On the permanence of stored information in the human brain. *American Psychologist, 35,* 409–420.

Luria, A. R. (1979). *The making of mind: A personal account of Soviet psychology.* Cambridge, MA: Harvard University Press.

Lynch, G., & Baudry, M. (1984). The biochemistry of memory: A new and specific hypothesis. *Science, 224,* 1057–1063.

MacFarlane, J. A. (1975). Olfaction in the development of social preferences in the human neonate. In M. A. Hofer (Ed.), *Parent-infant interaction.* New York: Elsevier.

MacKenzie, B. (1984). Explaining race differences in IQ. *American Psychologist, 39,* 1214-1233.

MacKinnon, D. W. (1962). The nature and nurture of creative talent. *American Psychologist, 17,* 484–495.

MacKinnon, D. W. (1978). *In search of human effectiveness.* Buffalo, NY: Creative Education Foundation.

MacLeod, J. (1987). *Ain't no making it: Leveled aspirations in a low-income neighborhood.* Boulder, CO: Westview.

Mahone, G. H. (1960). Fear of failure and unrealistic vocational aspirations. *Journal of Abnormal and Social Psychology, 47,* 166–175.

Mandler, G., & Sarason, S. B. (1952). A study of anxiety and learning. *Journal of Abnormal and Social Psychology, 47,* 166–173.

Manns, W. (1981). Support systems and significant others in black families. In H. P. McAdoo (Ed.), *Black families.* Beverly Hills, CA: Sage.

Marx, K. (1906). *Capital* (Vol. 1). New York: Modern Library. (Original work published 1867)

Maslow, A. (1970). *Motivation and personality* (2nd ed.). New York: Harper & Row.

Massey, D. S., & Bitterman, B. (1985). Explaining the paradox of Puerto Rican segregation. *Social Force, 64,* 306–331.

Massey, D. S., & Denton, N. A. (1989). Hypersegregation in U.S. metropolitan areas: Black and Hispanic segregation along five dimensions. *Demography, 26,* 373–391.

Masters, W. H., & Johnson, V. E. (1970). *Human sexual inadequacies.* New York: Little Brown.

Maurer, D., & Salapatek, P. (1976). Developmental changes in the scanning of faces by young infants. *Child Development, 47,* 523–527.

McAuther, L. Z. (1972). The how and what of why: Some determinants and consequences of causal attribution. *Journal of Personality and Social Psychology, 22,* 171–193.

McAuthur, L. A. (1970). Luck is alive and well in New Haven. *Journal of Personality and Social Psychology, 16,* 316–318.

McClelland, D. C. (1958). Risk-taking in children with high and low need for achievement. In J. W. Atkinson (Ed.), *Motives in fantasy, action, and society.* Princeton, NJ: Van Nostrand.

McClelland, D. C. (1961). *The achieving society.* Princeton, NJ: Van Nostrand.

McClelland, D. C. (1978). Managing motivation to expand human freedom. *American Psychologist, 33,* 201–210.

McClelland, D. C., & Atkinson, J. W. (1948). The projective expression of needs: I. The effect of ego involvement, success and failure on perception. *Journal of Psychology, 27,* 311–330.

McClelland, D. C., Atkinson, J. W., Clark, R. A., & Lowell, E. L. (1953). *The achievement motive.* New York: Appleton-Century-Crofts.

McClelland, D. C., Clark, R. A., Roby, T. B., & Atkinson, J. W. (1949). The projective expression of need for achievement on thematic apperception. *Journal of Experimental Psychology, 39,* 242–255.

McClelland, D. C., & Friedman, G. A. (1952). A cross-cultural study of the relationship between child-training practices and achievement motivation appearing in folk tales. In G. E. Swanson, T. M. Newcomb, & E. L. Hartley (Eds.), *Readings in social psychology,* New York: Holt.

McClelland, D. C., Rindlisbacher, A., & deCharms, R. C. (1955). Religious and other sources of parental attitudes toward independence training. In D. C. McClelland (Ed.), *Studies in motivation.* New York: Appleton-Century-Crofts.

McCullough, W. S., & Pitts, W. H. (1943). A logical calculus of the ideas immanent in nervous activity. *Bulletin of Mathematical Biophysics, 5,* 115.

McGinnies, E., Nordholm, L. A., Ward, C. D., & Bhanthumnauin, D. L. (1974). Sex and cultural differences in perceived locus of control among students in five countries. *Journal of Consulting and Clinical Psychology, 42,* 451–455.

McKeithen, K. B., Reitman, J. S., Rueter, H. H., & Hirtle, S. C. (1981). Knowledge organization and skill differences in computer programmers. *Cognitive Psychology, 13,* 307–325.

Mead, G. H. (1962). *Mind, self and society.* Chicago, IL: University of Chicago Press. (Original work published 1934)

Medin, D. L., & Cole, M. (1975). Comparative psychology and human cognition. In W. K. Estes (Ed.), *Handbook of learning and cognitive processes.* Hillsdale, NJ: Erlbaum.

Meichenbaum, D. H., & Goodman, J. (1971). Training impulsive children to talk to themselves: A means of developing self-control. *Journal of Abnormal Psychology, 77,* 115–122.

Merton, R. (1946). *Mass persuasion.* New York: Harper.

Messer, S. B. (1976). Reflection-impulsivity: A review. *Psychological Bulletin, 83,* 1026–1052.

Mill, J. S. (1949). *A system of logic.* London: Longmans, Green & Company. (Original work published 1843)

Miller, D. T., & Ross, M. (1975). Self-serving biases in attribution of causality: Fact or fiction? *Psychological Bulletin, 82,* 213–225.

Miller, G. (1956). The magical number seven, plus or minus two: Some limits on our capacity for processing information. *Psychological Review, 63,* 81–97.

Miller, J., Schooler, C., Kohn, M. L., & Miller, K. A. (1979). Women and work: The psychological effects of occupational conditions. *American Journal of Sociology, 85,* 66–94.

Miller, J., Slomczynski, K. M., Kohn, M. L. (1987). Continuity of learning-generalization through the life span: The effect of job on men's intellectual process in the United States and Poland. In C. Schooler & K. W. Schaie (Eds.), *Cognitive functioning and social structure over the life course.* Norwood, NJ: Ablex.

Miller, K. A., & Kohn, M. L. (1983). The reciprocal effects of job conditions and the intellectuality of leisure-time activities. In M. L. Kohn & C. Schooler (Eds.), *Work and personality: An inquiry into the impact of social stratification.* Norwood, NJ: Ablex.

Minton, H. L. (1988). *Lewis M. Terman: Pioneer in psychological testing.* New York: New York University Press.

Mirels, H. L. (1970). Dimensions of internal versus external control. *Journal of Consulting and Clinical Psychology, 34,* 226–228.

Mischel, W., Zeiss, R., & Zeiss, A. (1974). An internal-external control test for young children. *Journal of Personality and Social Psychology, 29,* 265–278.

Monahan, L., Kuhn, M., & Shaver, P. (1974). Intrapsychic versus cultural explanations of the "fear of success" motive. *Journal of Personality and Social Psychology, 29,* 60–64.

Montgomery, K. C. (1954). The role of exploratory drive in learning. *Journal of Comparative and Physiological Psychology, 47,* 60–64.

Moore, L. L. (1972). The relationship of academic group membership to the motive to avoid success in women. *Dissertation Abstract International, 32,* 4355.

Morgan, H. H. (1952). A psychometric comparison of achieving and non-achieving college students of high ability. *Journal of Consulting Psychology, 16,* 292–298.

Morris, J. L. (1966). Propensity for risk taking as determinant of vocational choice. *Journal of Personality and Social Psychology, 3,* 328–335.

Moscovite, I., & Craig, W. J. (1983). *The impact of federal cutbacks on working AFDC recipients in Minnesota.* Minneapolis, MN: University of Minnesota, Center for Health Services Research.

Moss, H. A., & Kagan, J. (1961). Stability of achievement and recognition seeking behaviors from early childhood through adulthood. *Journal of Abnormal and Social Psychology, 52,* 504–513.

Motte, E. (1979). Thérapeutique chez les Pygmées Aka de Mongoumba. In S. Bahuchet (Ed.), *Pygmées de Centrafrique.* Paris: Société d'Etudes Linguistiques et Anthropologiques de France.

Moulton, R. W. (1965). Effects of success and failure on level of aspiration as related to achievement motives. *Journal of Personality and Social Psychology, 1,* 399–406.

Murphy, L. B. (1974). Coping, vulnerability, and resilience in childhood. In G. V. Coelho, D. A. Hamburg, & J. E. Adams (Eds.), *Coping and adaptation.* New York: Basic Books.

Murray, H. A. (1938). *Explorations in personality.* New York: Oxford University Press.

Murray, L., & Trevarthen, C. (1985). Emotional regulation of interactions between two-month-olds and their mothers. In T. Field & N. Fox (Eds.), *Social perception in infants.* Norwood, NJ: Ablex.

Naditch, M. P., & de Maio, D. (1975). Locus of control and competence. *Journal of Personality, 43,* 541–559.

Naisbitt, J. (1984). *Megatrends.* New York: Warner Books.

Nakazima, S. (1962). A comparative study of the speech developments of Japanese and American English in children. *Studies in Phonology, 2,* 27–39.

National Commission on Excellence. (1983). *A nation at risk.* Washington, DC: Author.

Neisser, A. (1983). *The other side of silence.* New York: Knopf.

Neisser, U. (1963). The imitation of man by machine. *Science, 139,* 193–197.

Neisser, U. (1967). *Cognitive psychology.* New York: Appleton-Century-Crofts.

Neisser, U. (1976a). *Cognition and reality: Principles and implications of cognitive psychology.* San Francisco, CA: Freeman.

Neisser, U. (1976b). General, academic, and artificial intelligence. In L. B. Resnick (Ed.), *The nature of intelligence.* Hillsdale, NJ: Erlbaum.

Neisser, U. (1978). The concept of intelligence. *Intelligence, 3,* 217–227.

Neisser, U. (1981). John Dean's memory: A case study. *Cognition, 9,* 1–22.

Neisser, U. (1982). *Memory observed: Remembering in natural context.* New York: Freeman.

Neisser, U. (1983). Components of intelligence or steps in routine procedures? *Cognition, 15,* 189–197.

Neisser, U. (1984). Interpreting Harry Bahrick's discovery: What confers immunity against forgetting? *Journal of Experimental Psychology: General, 113,* 32–35.

Neisser, U. (1986a). Nested structure in autobiographical memory. In D. C. Rubin (Ed.), *Autobiographical memory.* New York: Cambridge University Press.

Neisser, U. (Ed.). (1986b). *The school achievement of minority children: New perspectives.* Hillsdale, NJ: Erlbaum.

Neisser, U. (1987). From direct perception to conceptual structure. In U. Neisser (Ed.), *Concepts and conceptual development: Ecological and intellectual factors in categorization.* New York: Cambridge University Press.

Neisser, U. (1988). What is ordinary memory the memory of? In U. Neisser & E. Winograd (Eds.), *Remembering reconsidered: Ecological and traditional approaches to the study of memory.* New York: Cambridge University Press.

Neisser, U. (1990). Direct perception and other forms of knowing. In R. R. Hoffman (Ed.), *Cognition and the symbolic process* (Vol. 3). Hillsdale, NJ: Erlbaum.

Nelson, K. (1973). Structure and strategy in learning to talk. *Monographs of the Society for Research on Child Development, 38* (Serial No. 149).

Nelson, K. (1974). Concept, word, and sentence: Interrelations in acquisition and development. *Psychological Review, 81,* 267–285.

Nelson, K. (1986). *Event knowledge: Structure and function in development.* Hillsdale, NJ: Erlbaum.

Nelson, K., & Ross, G. (1980). The generalities and specifics of long-term memory in infants and young children. In M. Perlmutter (Ed.), *New directions for child development.* San Francisco: Jossey-Bass.

Newell, A. (1980). Reasoning, problem solving, and decision processes: The problem space hypothesis. In R. Nickerson (Ed.), *Attention and performance: VIII.* Hillsdale, NJ: Erlbaum.

Nicholls, J. G. (1975). Causal attributions and other achievement related cognitions: Effects of task outcome, attainment value and sex. *Journal of Personality and Social Psychology, 31,* 379–389.

Nicholls, J. G., & Miller, A. T. (1984). Development and its discontents: The differentiation of the concept of ability. *Advances in Motivation and Achievement, 3,* 185–218.

Nowicki, S. (1978). Reported stressful events during developmental periods and their relation to locus of control orientation in college students. *Journal of Consulting and Clinical Psychology, 6,* 1552–1553.

Nowicki, S., & Duke, M. P. (1974). A preschool and primary locus of control scale. *Developmental Psychology, 10,* 874–880.

Nowicki, S., & Duke, M. P. (1983). The Nowicki-Strickland life-span locus of control scale: Construct validation. In H. M. Lefcourt (Ed.), *Research with the locus of control construct* (Vol. 2). New York: Academic.

Nowicki, S., & Strickland, B. R. (1973). A locus of control scale for children. *Journal of Consulting Psychology, 40,* 148–154.

Ogbu, J. U. (1985). A cultural ecology of competence among inner-city blacks. In M. B. Spencer, G. K. Brookins, & W. R. Allen (Eds.), *Beginnings: The social and affective development of black children.* Hillsdale, NJ: Erlbaum.

Ogbu, J. U. (1986). The consequences of the American caste system. In U. Neisser (Ed.), *The school achievement of minority children: New perspectives.* Hillsdale, NJ: Erlbaum.

Oliver, R. R., & Hornsby, J. R. (1966). On equivalence. In J. S. Bruner, R. R. Oliver, & P. M. Greenfield (Eds.), *Studies in cognitive growth.* New York: Wiley.

Ornstein, P. A., Naus, M. J., & Stone, B. P. (1977). Rehearsal training and developmental differences in memory. *Developmental Psychology, 13,* 15–24.

Osborn, A. (1957). *Applied imagination.* New York: Scribner.

Otten, M. W. (1977). Inventory and expressive measures of locus of control and academic performance: A five year outcome study. *Journal of Personality Assessment, 43,* 401–405.

Owens, W. A. (1966). Age and mental abilities: A second adult follow-up. *Journal of Educational Psychology, 57,* 311–325.

Padilla, A. M., Lindholm, K. J., Chen, A., Duran, R., Hakuta, K., Lambert, W., & Tucker, G. R. (1991). The English-only movement: Myths, reality, and implications for psychology. *American Psychologist, 46,* 120–130.

Palincsar, A. S., & Brown, A. L. (1984). Reciprocal teaching of comprehension—fostering and monitoring activities. *Cognition and Instruction, 1,* 117–175.

Papert, S. (1980). *Mind-storms: Children, computers, and powerful ideas.* New York: Basic Books.

Parham, W. D., & Parham, T. A. (1989). The community and academic achievement. In G. L. Berry & J. K. Asamen (Eds.), *Black students: Psychological issues and academic achievement.* Newbury Park, CA: Sage.

Paris, S. G., & Oka, E. R. (1986). Children's reading strategies, metacognition, and motivation. *Developmental Review, 6,* 25–56.

Parsons, J. E., Adler, T. F., & Kaczals, C. M. (1982). Socialization of achievement attitudes and beliefs: Parental influence. *Child Development, 53,* 310–321.

Peirce, C. S. (1955). *Philosophical writings of Peirce* (J. Buchler, Ed.). New York: Dover. (Original articles published around 1878)

Pellegrino, J. W., & Glaser, R. (1979). Cognitive components and correlates in the analysis of individual differences. *Intelligence, 3,* 187–214.

Pelto, P. J. (1968, April). The differences between "tight" and "loose" societies. *Transaction,* 37–40.

Penfield, W. (1975). *The mystery of the mind: A critical study of consciousness and the human brain.* Princeton, NJ: Princeton University Press.

Perkins, D. N. (1986). *Knowledge as design.* Hillsdale, NJ: Erlbaum.

Peterson, C., & Seligman, M. E. P. (1984). Causal explanations as a risk factor for depression: Theory and evidence. *Psychological Review, 91,* 347–374.

Petroni, F. A. (1970). "Uncle Tom": White stereotypes in the black movement. *Human Organization, 29,* 260–266.

Phares, E. J. (1968). Differential utilization of information as a function of internal-external control. *Journal of Personality, 36,* 649–662.

Phares, E. J. (1973). *Locus of control: A personality determinant of behavior.* Morristown, NJ: General Learning Press.

Phares, E. J. (1976). *Locus of control in personality.* Morristown, NJ: General Learning Press.

Piaget, J. (1952). *The origins of intelligence in children.* (M. Cook, Trans.). New York: International Universities Press.

Piaget, J., & Inhelder, B. (1973). *Memory and intelligence.* New York: Basic Books.

Picott, R. (1976). *A quarter century of elementary and secondary education.* Washington, DC: Association for the Study of Negro Life and History.

Piedmont, R. L. (1988). An interactional model of achievement motivation and fear of success. *Sex Roles, 19,* 467–490.

Pinkney, A. (1984). *The myth of black progress.* Cambridge, England: Cambridge University Press.

Plant, W. T. (1965). Personality changes associated with college attendance. *Human Development, 8,* 142–151.

Plomin, R., Defries, J. C., & McClearn, G. E. (1980). *Behavioral genetics: A primer.* San Francisco, CA: Freeman.

Pottharst, B. C. (1955). *The achievement motive and level of aspiration after experimentally induced success and failure.* Unpublished doctoral dissertation, University of Michigan, Ann Arbor.

Pouissant, A. F. (1974). Building a strong self-image in the black child. *Ebony, 29,* 138–143.

Pressley, M., Borkowski, J. C., & O'Sullivan, J. T. (1984). Memory strategy instruction is made of this: Metamemory and durable strategy use. *Educational Psychologist, 19,* 84–107.

Pressley, M., Borkowski, J. G., & O'Sullivan, J. T. (1985). Children's metamemory and the teaching of memory strategies. In D. L. Forrest-Pressley, G. E. MacKinnon, & T. G. Waller (Eds.), *Metacognition, cognition, and human performance* (Vol. 1). New York: Academic.

Pressley, M., & Levin, J. R. (1977). Developmental differences in subjects' associative-learning strategies and performance: Assessing a hypothesis. *Journal of Experimental Child Psychology, 24,* 431–439.

Prociuk, T. J., & Breen, L. J. (1975). Defensive externality and academic performance. *Journal of Personality and Social Psychology, 31,* 549–556.

Proshanky, H., & Newton, P. (1968). The nature and meaning of Negro self-identity. In M. Deutsch, I. Katz, & A. R. Jensen (Eds.), *Social class, race, and psychological development.* New York: Holt, Rinehart and Winston.

Putnam, H. (1971). *Philosophy of logic.* New York: Harper & Row.

Quillian, M. R. (1968). Semantic memory. In M. Minsky (Ed.), *Semantic information processing.* Cambridge: Cambridge University Press.

Rapaczynski, W., Singer, D. G., & Singer, J. L. (1982). Teaching television: A curriculum for young children. *Journal of communication, 2,* 46–55.

Raynor, J. O. (1974). Future orientation in the study of achievement motivation. In J. W. Atkinson & J. O. Raynor (Eds.), *Motivation and achievement.* Washington, DC: Winston.

Reese, H. W. (1962). Verbal mediation as a function of age level. *Psychological Bulletin, 59,* 502–509.

Regian, J., Shute, V., & Pellegrino, J. (1985, November). *The modifiability of spatial processing skills.* Paper presented at the 26th meeting of the Psychonomic Society, Boston, MA.

Reid, D. W., & Ziegler, M. (1981). The desired control measure and adjustment among the elderly. In H. M. Lefcourt (Ed.), *Research with the locus of control construct* (Vol. 1). New York: Academic.

Reid, M. K., & Borkowski, J. G. (1987). Causal attributions of hyperactive children: Implications for training strategies and self-control. *Journal of Educational Psychology, 79,* 296–307.

Remanis, G. (1971, July). *Effects of experimental IE modification techniques and home environmental variables on IE.* Paper presented at the American Psychological Association Convention, Washington, DC.

Resnick, L. B. (1983). A developmental theory of number understanding. In H. P. Ginsburg (Ed.), *The development of mathematical thinking.* New York: Academic.

Resnick, L. B., & Glaser, R. (1976). Problem solving and intelligence. In L. B. Resnick (Ed.), *The nature of intelligence.* Hillsdale, NJ: Erlbaum.

Rhoewalt, F., & Davison, J. (1984). *Self-handicapping and subsequent performance: The role of outcome valance and attributional certainty.* Unpublished manuscript, University of Utah, Salt Lake City.

Rholes, W. S., Blackwell, J., Jordan, C., & Walters, C. (1980). A developmental study of learned helplessness. *Developmental Psychology, 16,* 616–624.

Ricciuti, H. N., & Sadacca, R. (1955). *The prediction of academic grades with a projective test of achievement motivation: II. Cross validation at the high school level.* Princeton, NJ: Educational Testing Service.

Rist, R. C. (1970). Student social class and teacher expectations: The self-fulfilling prophecy and ghetto education. *Harvard Educational Review, 40,* 411–451.

Robbins, L., & Robbins, E. (1973). Comment on "Toward an understanding of achievement-related conflicts in women." *Journal of Social Issues, 29,* 133–137.

Roberts, K., & Horowitz, F. D. (1986). Basic level categorization in seven- and nine-month-old infants. *Journal of Child Language, 13,* 191–206.

Rodin, J., & Langer, E. J. (1977). Long-term effects of a control-relevant intervention with the institutional aged. *Journal of Personality and Social Psychology, 35,* 897–902.

Rogers, C. R. (Ed.). (1967). *The therapeutic relationship and its impact: A study of psychotherapy with schizophrenics.* Madison, WI: University of Wisconsin Press.

Rogoff, B. (1990). *Apprenticeship in thinking: Cognitive development in social context.* New York: Oxford University Press.

Rogoff, B., & Waddell, K. J. (1982). Memory for information organization in a scene by children from two cultures. *Child Development, 53,* 1224–1228.

Rohrkemper, M., & Bernshon, B. L. (1983). The quality of student task engagement: Elementary school students' reports of the causes and effects of problem difficulty. *Elementary School Journal, 85,* 127–147.

Rokeach, M. (1960). *The open and closed mind.* New York: Knopf.

Rosch, E. (1973). Natural categories. *Cognitive Psychology, 4,* 328–350.

Rosch, E. (1974). Linguistic relativity. In A. Silverstein (Ed.), *Human communication.* Hillsdale, NJ: Erlbaum.

Rosch, E. (1978). Principles of categorization. In E. Rosch & B. L. Lloyd (Eds.), *Cognition and categorization.* Hillsdale, NJ: Erlbaum.

Rosch, E., Mervis, C. B., Gray, W. D., Johnson, D. M., & Boyes-Braem, P. (1976). Basic objects in natural categories. *Cognitive Psychology, 8,* 382–439.

Rosen, B. C. (1959). Race, ethnicity, and the achievement syndrome. *American Sociological Review, 24,* 47–60.

Rosen, B. C., & D'Andrade, R. G. (1959). The psychosocial origins of achievement motivation. *Sociometry, 22,* 185–218.

Rosenbaum, J. (1976). *Making inequality: The hidden curriculum of high school tracking.* New York: Wiley.

Rosenberg, M. (1965). *Society and the adolescent self-image.* Princeton, NJ: Princeton University Press.

Rosenberg, M. (1979). *Conceiving the self.* New York: Basic Books.

Rosenberg, M. (1981). The self-concept: Social product and social force. In M. Rosenberg & R. H. Turner (Eds.), *Social psychology: Sociological perspectives.* New York: Basic Books.

Rosenberg, M., & Pearlin, L. I. (1978). Social class and self-esteem among children and adults. *American Journal of Sociology, 84,* 53–77.

Rosenberg, M., & Simmons, R. G. (1972). *Black and white self-esteem: The urban school child.* Washington, DC: American Sociological Association.

Rosenzweig, M. R. (1984). Experience, memory, and the brain. *American Psychologist, 39,* 365–376.

Rosenzweig, M. R., Bennett, E. L., & Diamond, M. C. (1972). Brain changes in response to experience. *Scientific American, 226,* 22–29.

Ross, M. (1976). The self-perception of intrinsic motivation. In J. H. Harvey, W. J. Ickes, & R. F. Kidd (Eds.), *New directions in attribution research.* Hillsdale, NJ: Erlbaum.

Rossi, P. H. (1990). The old homeless and the new homeless in historical perspective. *American Psychologist, 45,* 954–959.

Rotter, D. M., Langland, L., & Berger, D. (1971). The validity of tests of creative thinking in seven-year-old children. *Gifted Child Quarterly, 43,* 471–480.

Rotter, J. B. (1954). *Social learning and clinical psychology.* New York: Prentice-Hall.

Rotter, J. B. (1966). Generalized expectancies for internal versus external control of reinforcement. *Psychological Monographs, 80* (Whole No. 609).

Rotter, J. B. (1975). Some problems and misconceptions related to the construct of internal versus external control of reinforcement. *Journal of Consulting and Clinical Psychology, 48,* 56–67.

Rotter, J. B., & Mulry, R. C. (1965). Internal versus external locus of reinforcement and decision time. *Journal of Personality and Social Psychology, 2,* 598–604.

Russell, B. (1917). *Mysticism and logic.* London: Unwin.

Rutter, M. (1978). Early sources of security and competence. In J. Bruner & A. Garton (Eds.), *Human growth and development.* New York: Oxford University Press.

Rutter, M. (1985). Resilience in the face of adversity: Protective factors and resistance to psychiatric disorder. *British Journal of Psychiatry, 147,* 598–611.

Salapatek, P. (1968). Visual scanning of geometric figures by the human newborn. *Journal of Comparative and Physiological Psychology, 66,* 247–258.

Salili, F., Maehr, M. L., & Gillmore, G. (1976). Achievement and mortality: A cross-cultural analysis of causal attribution and evaluation. *Journal of Personality and Social Psychology, 33,* 327–337.

Salomon, G. (1981). *Communication and education: Social and psychological interactions.* Beverly Hills, CA: Sage.

Salomon, G. (1984). Television is "easy" and print is "tough": The differential investment of mental effort in learning as a function of perception and attributions. *Journal of Educational Psychology, 76,* 647–658.

Sanders-Phillips, K. (1989). Prenatal and postnatal influences on cognitive development. In G. L. Berry & J. K. Asamen (Eds.), *Black students: Psychosocial issues and academic achievement.* Newbury Park, CA: Sage.

Sapir, E. (1921). *Language: An introduction to the study of speech.* New York: Harcourt.

Sartre, J. P. (1963). *The psychology of imagination.* New York: Citadel. (Original work published 1936)

Saxe, G. B. (1981). Body parts as numerals: A developmental analysis of numeration among the Oksapmin in Papua, New Guinea. *Child Development, 52,* 306–316.

Saxe, G. B., & Posner, J. K. (1983). The development of numerical cognition: Cross-cultural perspectives. In H. P. Ginsburg (Ed.), *The development of mathematical thinking.* New York: Academic.

Scarr, S., Pakstis, A. J., Katz, S. H., & Barker, W. B. (1977). The absence of a relationship between degree of white ancestry and intellectual skills within a black population. *Human Genetics, 39,* 69–86.

Scarr, S., & Weinberg, R. A. (1976). IQ test performance of black children adopted by white families. *American Psychologist, 31,* 726–739.

Schactel, E. G. (1947). On memories and childhood amnesia. *Psychiatry, 10,* 1–26.

Schank, R. C., & Abelson, R. P. (1977). *Scripts, plans, goals and understanding.* Hillsdale, NJ: Erlbaum.

Schneider, W. (1985). Developmental trends in the metamemory-memory behavior relationship: An integrative review. In D. L. Forrest-Pressley, G. E. MacKinnon,

& T. G. Waller (Eds.), *Metacognition, cognition, and human performance* (Vol. 1). New York: Academic.

Schneider, W., Borkowski, J. G., Kurtz, B. E., & Kerwin, K. (1986). Metamemory and motivation: A comparison of strategy use and performance in German and American children. *Journal of Cross-cultural Psychology, 17,* 315–336.

Schneider, W., Korkel, J., & Weinert, F. E. (1990). Expert knowledge and general abilities and test processing. In W. Schneider & F. Weinert (Eds.), *Interactions among aptitudes, strategies, and knowledge in cognitive performance.* New York: Springer-Verlag.

Schneider, W., & Weinert, F. (Eds.). (1990). *Interactions among attitudes, strategies, and knowledge in cognitive performance.* New York: Springer-Verlag.

Scribner, S. (1984). Studying working intelligence. In B. Rogoff & J. Lave (Eds.), *Everyday cognition.* Cambridge, MA: Harvard University Press.

Scribner, S. (1986). Thinking in action: Some characteristics of practical thought. In R. J. Sternberg & R. K. Wagner (Eds.), *Practical intelligence: Nature and origins of competence in the everyday world.* New York: Cambridge University Press.

Scribner, S., & Cole, M. (1973). Cognitive consequence of formal and informal education. *Science, 182,* 553–559.

Scribner, S., & Cole, M. (1981). *The psychology of literacy.* Cambridge, MA: Harvard University Press.

Schultz, R. (1976). Efforts of control and predictability on the physical and psychological well-being of the institutionalized aged. *Journal of Personality and Social Psychology, 33,* 563–573.

Schwartz, J. T. (1988). The new connectionism: Developing relationships between neuroscience and artificial intelligence. In S. R. Graubard (Ed.), *The artificial intelligence debate: False starts, real foundations.* Cambridge, MA: MIT Press.

Seeman, M. (1963). Alienation and social learning in a reformatory. *American Journal of Sociology, 69,* 270–284.

Seeman, M., & Evans, J. W. (1962). Alienation and learning in a hospital setting. *American Sociological Review, 27,* 772–783.

Seligman, M. E. P. (1975). *Helplessness: On depression, development, and death.* San Francisco, CA: Freeman.

Sells, L. W. (1978). Mathematics—A critical filter. *The Science Teacher, 45,* 28–29.

Sennett, R., & Cobb, J. (1972). *The hidden injuries of class.* New York: Vintage.

Seyfried, B. A., & Hendrick, C. (1973). When do opposites attract? When they are opposite in sex and sex-role attitudes. *Journal of Personality and Social Psychology, 25,* 15–20.

Shaffer, D. R., & Wegley, C. (1974). Success orientation and sex-role congruence as determinants of the attractiveness of competent women. *Journal of Personality, 42,* 586–600.

Shatz, M., & Gelman, R. (1973). The development of communication skills: Modifications in the speech of young children as a function of listener. *Monographs of the Society for Research in Child Development, 38* (Serial No. 152).

Sheingold, K. (1973). Developmental differences in intake and storage of visual information. *Journal of Experimental Child Psychology, 16,* 1–11.

Sherif, M., & Sherif, C. W. (1953). *Groups in harmony and tension.* New York: Harper & Row.

Sherman, M. R., Pelletier, R. J., & Ryckman, R. M. (1973). Replication of the relationship between dogmatism and locus of control. *Psychological Reports, 33,* 749–750.

Sherwood, J. (1965). Self identity and referent others. *Sociometry, 28,* 66–81.

Shore, R. E. (1967). *Parental determinants of boys' internal-external control.* Unpublished doctoral dissertation, Syracuse University, Syracuse, NY.

Siegler, R. S. (1987). Strategy choices in subtraction. In J. A. Sloboda & D. Rogers (Eds.), *Cognitive processes in mathematics.* Oxford: Clarendon.

Simon, D. P., & Simon, H. A. (1978). Individual differences in solving physics problems. In R. S. Siegler (Ed.), *Children's thinking: What develops?* Hillsdale, NJ: Erlbaum.

Simon, H. A. (1974). How big is a chunk? *Science, 183,* 482–489.

Simonton, D. K. (1984). *Genius, creativity and leadership.* Cambridge, MA: Harvard University Press.

Singer, J. L., & Singer, D. G. (1983). Psychologists look at television: Cognitive, developmental, personality and social policy implications. *American Psychologist, 38,* 826–834.

Singer, J., Singer, D., & Rapaczynski, W. (1984). Family patterns and television viewing as predictors of children's beliefs and aggression. *Journal of Communication, 34,* 73–89.

Slaughter, D. T. (1986, April). *Children's peer acceptance and parental involvement in desegregated private elementary schools.* Paper presented at the meeting of the American Educational Research Association, San Francisco, CA.

Slaughter, D. T., & McWorter, G. A. (1985). Social origins and early features of the scientific study of black American families and children. In M. B. Spencer, G. K. Brookins, & W. R. Allen (Eds.), *Beginnings: The social and affective development of black children.* Hillsdale, NJ: Erlbaum.

Slavin, R. E. (1982). *Cooperative learning: Student teams.* Washington, DC: National Education Association.

Slavin, R. E. (1987). Cooperative learning and the cooperative school. In D. S. Strickland & E. L. Cooper (Eds.), *Educating black children: America's challenge.* Washington, DC: Howard University Press.

Slavin, R. E., & Madden, N. A. (1979). School practices that improve race relations. *American Educational Research Journal, 16,* 169–180.

Slobin, D. I. (1970). Universals of grammatical development in children. In G. B. Flores, J. Arcais, & W. J. M. Levelt (Eds.), *Advances in psycholinguistics.* Amsterdam: North-Holland Publishing.

Smith, A. (1937). *An inquiry into the nature and cause of the wealth of nations.* New York: Random House. (Original work published 1776)

Smith, E. J. (1982). The black female adolescent: A review of the educational, career and psychological literature. *Psychology of Women Quarterly, 6,* 261–288.

Smith, J. P. (1988). Tomorrow's white teacher: A response to the Holmes Group. *Journal of Negro Education, 57,* 178–194.

Snyder, M. L., Smoller, B., Strenta, A., & Frankel, A. (1981). A comparison of egotism, negativity, and learned helplessness as explanations for poor performance after unsolvable problems. *Journal of Personality and Social Psychology, 40,* 24–30.

Sodian, B., Schneider, W., & Perlmutter, M. (1986). Recall, clustering, and metamemory in young children. *Journal of Experimental Child Psychology, 41,* 395–410.

Sorrentino, R. M., & Short, J. A. (1974). Effects of fear of success on women's performance at masculine versus feminine tasks. *Journal of Research in Personality, 8,* 277–290.

Spearman, C. (1904). General intelligence objectively determined and measured. *American Journal of Psychology, 15,* 201–293.

Spearman, C. (1927). *The ability of man.* New York: Macmillan.

Spence, A. J., & Spence, S. H. (1980). Cognitive changes associated with social skills training. *Behavioral Research and Therapy, 18,* 265–272.

Spence, J. T., & Helmreich, R. L. (1978). *Masculinity and femininity: Their psychological dimensions, correlates, and antecedents.* Austin, TX: University of Texas Press.

Spence, J. T., & Helmreich, R. L. (1983). Achievement-related motives and behavior. In J. T. Spence (Ed.), *Achievement and achievement motives: Psychological and sociological approaches.* San Francisco, CA: Freeman.

Sperling, G. (1960). The information available in brief visual presentations. *Psychological Monographs, 74* (11, Whole No. 498).

Spurlock, J. (1973). Some consequences of racism for children. In C. V. Willie, B. M. Kramer, & B. S. Brown (Eds.), *Racism and mental health.* Pittsburgh, PA: University of Pittsburgh Press.

Squire, L. R. (1987). *Memory and brain.* New York: Oxford University Press.

Staszewski, J. J. (1990). Exceptional memory: The influence of practice and knowledge on the development of elaborative encoding strategies. In W. Schneider & F. E. Weinert (Eds.), *Interactions among aptitudes, strategies, and knowledge in cognitive performance.* New York: Springer-Verlag.

Steinmann, A., & Fox, D. J. (1970). Attitudes towards women's family role among black and white undergraduates. *Family Coordinator, 19,* 363–368.

Stephens, M. W. & Delys, P. (1973). External control expectancies among disadvantaged children at preschool age. *Child Development, 44,* 670–674.

Sternberg, R. J. (1977). *Intelligence, information processing and analogical reasoning: The componential analysis of human abilities.* Hillsdale, NJ: Erlbaum.

Sternberg, R. J. (1981). Intelligence and nonentrenchment. *Journal of Educational Psychology, 73,* 1–16.

Sternberg, R. J. (Ed.). (1984). *Advances in the psychology of human intelligence.* Hillsdale, NJ: Erlbaum.

Sternberg, R. J. (1985a). *Beyond IQ: A triarchic theory of human intelligence.* New York: Cambridge University Press.

Sternberg, R. J. (Ed.). (1985b). *Human abilities: An information processing approach.* New York: Freeman.

Sternberg, R. J. (1988). A three-facet model of creativity. In R. J. Sternberg (Ed.), *The nature of creativity: Contemporary psychological perspectives.* New York: Cambridge University Press.

Sternberg, R. J., Conway, B. E., Ketron, J. L., & Berstein, M. (1981). People's conception of intelligence. *Journal of Personality and Social Psychology, 41,* 37–55.

Sternberg, R. J., & Davidson, J. E. (1982). The mind of the puzzler. *Psychology Today, 16,* 37–44.

Sternberg, R. J., & Ditterman, D. K. (Eds.). (1986). *What is intelligence?: Contemporary viewpoints on its nature and definition.* Norwood, NJ: Ablex.

Sternberg, R. J., & Salter, W. (1982). Conceptions of intelligence. In R. J. Sternberg (Ed.), *Handbook of human intelligence.* New York: Cambridge University Press.

Stevenson, H. W., Lee, S. Y., & Stigler, J. W. (1986). Mathematics achievement of Chinese, Japanese, and American children. *Science, 231,* 693–699.

Steward, A. J., & Winter, D. G. (1974). Self-definition and social definition in women. *Journal of Personality, 42,* 238–259.

Stewart, L. H. (1977). Birth order and political leadership. In M. G. Hermann (Ed.), *The psychological examination of political leaders.* New York: Free Press.

Stillings, N. A., Feinstein, M. H., Garfield, J. L., Rissland, E. L., Rosenbaum, D. A., Weisler, S. E., & Baker-Ward, L. (1987). *Cognitive science: An introduction.* Cambridge, MA: MIT Press.

St. John, N. (1975). *School desegregation: Outcomes for children.* New York: Wiley.

Strickland, B. R. (1978). I-E expectations and health-related behavior. *Journal of Consulting and Clinical Psychology, 46,* 1192–1211.

Strickland, B. R. (1989). Internal-external control expectancies: From contingency to creativity. *American Psychologist, 44,* 1–12.

Sue, S., & Okazaki, S. (1990). Asian-American educational achievements: A phenomenon in search of an explanation. *American Psychologist, 45,* 913–920.

Sweeney, P. D., Anderson, K., & Bailey, S. (1986). Attributional style in depression: A meta-analysis review. *Journal of Personality and Social Psychology, 50,* 974–991.

Taylor, I. A. (1975). A retrospective view of creative investigation. In I. A. Taylor & J. W. Getzels (Eds.), *Perspectives in creativity.* Chicago, IL: Aldine.

Taylor, J. A. (1953). A personality scale of manifest anxiety. *Journal of Abnormal and Social Psychology, 48,* 285–290.

Taynor, J., & Deaux, K. (1973). When women are more deserving than men: Equity, attribution, and perceived sex differences. *Journal of Personality and Social Psychology, 28,* 360–367.

Teevan, R. C., & McGhee, P. E. (1972). Childhood development of fear of failure motivation. *Journal of Personality and Social Psychology, 21,* 345–348.

Terman, L. M. (1916). *The measure of intelligence.* Boston, MA: Houghton Mifflin.

Thomas, E. L. (1968). Movements of the eye. *Scientific American, 219,* 88–95.

Thompson, C. P., & Cowan, T. (1986). Flashbulb memories: A nicer interpretation of a Neisser recollection. *Cognition, 22,* 199–200.

Thurstone, L. L. (1938). *Primary mental abilities.* Chicago, IL: University of Chicago Press.

Torrance, E. P. (1972). Predictive validity of the Torrance tests of creative thinking. *Journal of Creative Behavior, 6,* 236–252.

Torrance, E. P. (1975). Creative research in education: Still alive. In I. A. Taylor & J. W. Getzels (Eds.), *Perspectives in creativity.* Chicago, IL: Aldine.

Torrance, E. P. (1979). *The search for Satori.* New York: Creative Educational Foundation.

Touhey, J. C. (1974). Effects of additional women professionals on ratings of occupational prestige and desirability. *Journal of Personality and Social Psychology, 29,* 86–90.

Travis, C. B., Wiley, D. L., McKenzie, B. J., & Kahn, A. S. (1983). *Sex and achievement domain: Cognitive patterns of success and failure.* Unpublished manuscript.

Trehub, S. E. (1976). The discrimination of foreign speech contrasts by infants and adults. *Child Development, 42,* 466–472.

Tucker, J. A., Vuchinich, R. E., & Sobell, M. B. (1981). Alcohol consumption as a self-handicapping strategy. *Journal of Abnormal Psychology, 90,* 220–230.

Tuddenham, R. D. (1948). Soldier intelligence in World Wars I and II. *American Psychologist, 3,* 54–56.

Tulkin, S. R., & Kagan, J. (1972). Mother-child interaction in the first year of life. *Child Development, 43,* 31–41.

Tulving, E. (1972). Episodic and semantic memory. In E. Tulving & W. Donaldson (Eds.), *Organization and memory.* New York: Academic.

Turing, A. M. (1937). On computable numbers, with an application to the Entscheidungsproblem. *Proceedings of the London Mathematical Society, 42,* 230–265.

Vernaud, G. (1982). A classification of cognitive tasks and operations of thought involved in addition and subtraction problems. In T. P. Carpenter, J. M. Moser, & T. A. Romberg (Eds.), *Addition and subtraction: A cognitive perspective.* Hillsdale, NJ: Erlbaum.

Vernon, P. E. (1969). *Intelligence and cultural enrichment.* London: Methuen.

Vygotsky, L. S. (1962). *Thought and language* (E. Haufmann & G. Vakar, Trans.). Cambridge, MA: MIT Press. (Original work published 1934)

Vygotsky, L. S. (1966). Play and its role in the mental development of the child. *Soviet Psychology, 5,* 6–18. (Original work published 1933)

Vygotsky, L. S. (1978). *Mind and society* (M. Cole, V. John-Steiner, S. Scribner, & E. Souberman, Eds. & Trans.). Cambridge, MA: Harvard University Press. (Original works published 1930–1935)

Wade, N. (1976). IQ and heredity: Suspicion of fraud beclouds classic experiment. *Science, 194,* 916–919.

Wagner, D. A. (1974). The development of short-term and incidental memory: A cross-cultural study. *Child Development, 48,* 389–396.

Wagner, D. A. (1978). Memories of Morocco: The influence of age, schooling, and environment on memory. *Cognitive Psychology, 10,* 1–28.

Wagner, R. K., & Sternberg, R. J. (1986). Tacit knowledge and intelligence in the everyday world. In R. J. Sternberg & R. K. Wagner (Eds.), *Practical intelligence: Nature and origins of competence in the everyday world.* New York: Cambridge University Press.

Walberg, H. J., Rasher, S. P., & Parkerson, J. (1980). Childhood and eminence. *Journal of Creative Behavior, 13,* 225–231.

Walker, C. H. (1987). Relative importance of domain knowledge and overall aptitude on acquisition of domain-related information. *Cognition and Instruction, 4,* 25–42.

Wallbrown, F. H., & Heulsman, C. D., Jr. (1975). The validity of the Wallach-Kogan creativity operations for inner-city children in two areas of visual art. *Journal of Personality, 43,* 109–126.

Walps, G. (1926). *The art of thought.* New York: Harcourt, Brace & World.

Waltz, D. L. (1988). The prospects for building truly intelligent machines. In S. R. Graubard (Ed.), *The artificial intelligence debate: False starts, real foundations.* Cambridge, MA: MIT Press.

Wason, P. C., and Johnson-Laird, P. N. (1972). *Psychology of reasoning: Structure and content.* Cambridge, MA: Harvard University Press.

Waters, H. S. (1982). Memory development in adolescence: Relationships between metamemory, strategy use, and performance. *Journal of Experimental Child Psychology, 33,* 183–195.

Weber, M. (1958). *The Protestant ethic and the spirit of capitalism* (T. Parson, Trans.). New York: Scribner. (Original work published 1904–05)

Webster, H. (1956). Some quantitative results. *Journal of Social Issues, 12,* 29–43.

Weinberg, M. (1977). *Minority students: A research appraisal.* Washington, DC: National Institute of Education, Department of Health, Education and Welfare.

Weiner, B., Frieze, I. H., Kukla, A., Reed, L., Rest, S., & Rosenbaum, R. M. (1972). Perceiving the causes of success and failure. In E. E. Jones, D. E. Kanouse, H. H. Kelley, R. E. Nisbett, S. Valins, & B. Weiner (Eds.), *Attributions: Perceiving the causes of behavior.* Morristown, NJ: General Learning Press.

Weiner, B., & Sierdo, J. (1975). Misattribution of failure and enhancement of achievement striving. *Journal of Personality and Social Psychology, 31,* 415–421.

Wellman, H. (1977). Tip of the tongue and feeling of knowing experiences: A developmental study of memory monitoring. *Child Development, 48,* 13–21.

Werker, J. F., Gilbert, J. H. V., Humphrey, K., & Tees, R. C. (1981). Developmental aspects of cross-language speech perception. *Child Development, 52,* 349–355.

Werner, E. E. (1986). Resilient offspring of alcoholics: A longitudinal study from birth to age 18. *Journal of Studies on Alcohol, 47,* 34–40.

Werner, H. (1948). *Comparative psychology of mental development.* New York: International Universities Press.

Wertheimer, M. (1961). Psychomotor co-ordination of auditory-visual space at birth. *Science, 134,* 1692.

Wertsch, J. V. (1985). *Vygotsky and the social formation of mind.* Cambridge, MA: Harvard University Press.

White, R. W. (1959). Motivation reconsidered: The concept of competence. *Psychological Review, 66,* 297–333.

Whorf, B. (1956). *Language, thought, and reality.* Cambridge, MA: MIT Press.

Wilson, W. J. (1987). *The truly disadvantaged: The inner city, the underclass, and public policy.* Chicago, IL: University of Chicago Press.

Wine, J. D. (1971). Test anxiety and direction of attention. *Psychological Bulletin, 76,* 92–104.

Wine, J. D. (1982). Evaluation anxiety: A cognitive-attentional construct. In H. W. Krohne & L. Laux (Eds.), *Achievement, stress, and anxiety.* Washington, DC: Hemisphere.

Winterbottom, M. (1958). The relation of childhood training in independence to achievement motivation. In J. W. Atkinson (Ed.), *Motives in fantasy, action and society.* Princeton, NJ: Van Nostrand.

Witkin, H. A., & Berry, J. W. (1975). Psychological differentiation in cross-cultural perspective. *Journal of Cross-Cultural Psychology, 6,* 4–87.

Witkin, H. A., & Goodenough, D. R. (1981). *Cognitive styles: Essence and origins.* New York: International Universities Press.

Witkin, H. A., Goodenough, D. R., & Karp, S. A. (1967). Stability of cognitive styles from childhood to young adulthood. *Journal of Personality and Social Psychology, 7,* 291–300.

Wittgenstein, L. (1953). *Philosophical investigations* (G. E. M. Anscombe, Trans.). Oxford: Blackwell.

Wittgenstein, L. (1961). *Tractatus logic-philosophicus* (D. F. Pears & B. F. McGuinness, Trans.). London: Routledge & Kegan Paul. (Original work published 1922)

Wolk, S., & DuCette, J. (1973). The moderating effect of locus of control in relation to achievement-motivation variables. *Journal of Personality, 41,* 59–70.

Woodward, K. L., & Denworth, L. (1990, May). The order of innovation: A study finds scientific rebels are born, not made. *Newsweek,* 76.

Yerkes, R. M. (Ed.). (1921). Psychological examining in the United States Army. *Memoirs of the National Academy of Science* (Vol. 15).

Yerkes, R. M., & Dodson, J. D. (1908). The relation of strengths of stimulus to rapidity of habit-formation. *Journal of Comparative Neurological Psychology, 18,* 459–482.

Younger, B. A., & Cohen, L. B. (1983). Infant perception of correlations among attributes. *Child Development, 54,* 858–867.

Yussen, S. R., & Levy, V. M., Jr. (1975). Developmental changes in predicting one's own span of short-term memory. *Journal of Experimental Child Psychology, 19,* 502–508.

Zajonc, R. B., & Markus, G. B. (1975). Birth order and intellectual development. *Psychological Review, 19,* 502–508.

Zajonc, R. B., Markus, H., & Markus, G. B. (1979). The birth order puzzle. *Journal of Personality and Social Psychology, 37,* 1325–1341.

Zuckerman, H. (1972). *Scientific elite.* New York: Free Press.

Zuckerman, M., & Wheeler, L. (1975). To dispel fantasies about fantasy-based measure of fear of success. *Psychological Bulletin, 82,* 932–946.

Authors Index

Subject Index